Water Current Turbines

A Fieldworker's Guide

Peter Garman

PRACTICAL ACTION
Publishing

Intermediate Technology Publications Ltd
trading as Practical Action Publishing
Schumacher Centre for Technology and Development
Bourton on Dunsmore, Rugby,
Warwickshire CV23 9QZ, UK
www.practicalactionpublishing.org

First published in 1986
Transferred to digital printing in 2008
Reprinted 2008

ISBN 978 0 946688 27 2

A catalogue record for this book is available from the British Library.

Since 1974, Practical Action Publishing has published and disseminated books
and information in support of international development work throughout
the world. Practical Action Publishing (formerly ITDG Publishing) is a trading
name of Intermediate Technology Publications Ltd (Company Reg. No. 1159018),
the wholly owned publishing company of Intermediate Technology
Development Group Ltd (working name Practical Action).
Practical Action Publishing trades only in support of its parent charity
objectives and any profits are covenanted back to Practical Action
(Charity Reg. No. 247257, Group VAT Registration No. 880 9924 76).

FOREWORD

The food crisis in Africa and elsewhere in the Third World highlights the need for new technologies which local communities can use to increase their agricultural production. An important means of achieving greater food production is by irrigation. Many arid areas are characterized by having large rivers or canals flowing through them. The volume of water is more than that needed to irrigate plots along or near to the banks; but traditional methods of water lifting are often inefficient, and modern methods are often too expensive.

This handbook describes the development of a new, simple and relatively inexpensive technology which, if used in the right circumstances, will lift water from the rivers on to the land. The water current turbine - you can think of it as a windmill inserted into the river current - has been tried and tested for three dry seasons at Juba on the White Nile, where it has been used profitably to irrigate small vegetable gardens.

We believe that our experience in the Sudan could provide the basis for extended trials of the turbine in other areas where similar conditions apply. We know that the turbine works; but before it is made freely available it is necessary to establish the social and economic circumstances within which it can be used by local people for their own benefit. After explaining the technical details, this handbook outlines the main socio-economic factors which must be taken into consideration before embarking upon a local project.

The purpose of the handbook is to inform development agencies and others of the availability of the technology, and to encourage them to test it out in their own circumstances. Mannufacturing drawings of two alternate designs are available from ITDG, and the author, Peter Garman, would be pleased to advise interested parties who wish to make and test the technology for themselves. Enquiries should be addressed to the author c/o The Information Office, ITDG, Myson House, Railway Terrace, Rugby, CV21 3HT, UK.

Dennis Frost
March 1986

iii

ACKNOWLEDGEMENTS

The work on which this book is based was funded by the Royal Netherlands Government, the original UK model turbine tests having been funded by the Hilden Trust. ITDG is very grateful for all the help these organisations gave. Thanks are due to many colleagues and friends at Juba Boatyard, ITDG and the Department of Engineering at Reading University for all their help and encouragement and to Mr Peter Giddens and Mr David Saunders for their advice on pumps.

I would like to thank Peter Burgess for writing the socio-economic analysis which forms Chapter 3 and for his considerable assistance in the completion of this book.

This book is dedicated to my workmates at Juba Boatyard.

Note: During the period covered by this book the Sudanese Pound was approximately equivalent to £0.54 Sterling but for the basis of the calculations used in Appendix 3 refer to page 112.

CONTENTS

CHAPTER 3 SOCIO-ECONOMIC ANALYSIS (by Peter Burgess)

LIST OF ILLUSTRATIONS

CHAPTER ONE

Introduction

1.1 THE OBJECTIVE OF THIS BOOK

This handbook is based on four years' experience of designing, building and field testing water current turbines (WCTs). Nine different turbines have been built and field tested, for a total of 15,500 running hours, at Juba on the White Nile. This experience has shown WCTs to be technically and economicaly viable as an alternative technology to small diesel pumps in southern Sudan (references 1 and 2 - see p 113.)

The assessment of whether a new technology is appropriate for a particular environment involves very many issues which can categorized under three broad headings:

1. is it technically operational - here?

2. is it economically attractive - here?

3. is it socially acceptable - here?

These three questions, and the multitude of more detailed issues raised by each, are interrelated and the different aspects require different emphasis in each situation.

The objective of this handbook is to draw on the operational experience gained in outhern Sudan to develop a guide to assist field workers in rural areas of poor countries in deciding whether they should investigate further the possibility of using water current turbines, and to offer a methodology for choosing between water current turbines and alternative small-scale water lifting devices. The aim here is to provide a checklist of the key physical, technical, economic and social factors relevant to assessing whether WCTs are the most appropriate technology in a given environment for a particular application.

In this book, irrigation is the main end-use of water which is considered. The reason for raising water has an important bearing on both the social acceptability and economic viability of the technology chosen. The focus here on irrigation is primarily because it offers the greatest potential as an economic activity through which WCTs can stimulate rural development. Other activities for which WCTs (in their present state of development) may have a role include:

(i)   raising water for livestock; and

(ii)   providing water for village industries.

Another possibility is pumping water for human consumption, but, in view of the health problems associated with this use of river water, the viability of WCTs for village water supply applications is not considered. Other potential uses are:

(i)   for electricity generation; and

(ii)   in direct mechanical applications.

However, WCT technology is not yet proven for these purposes and in consequence they are not discussed any further here.

1.2   Water Current Turbine (WCT) Technology

WCT Technology is described in detail in Chapter 2. As a result of the field experience at Juba, two systems have been developed:

(i)   The 'Mark 1' machine (swept area up to 5 square metres, shown in Figure 2.4) which depending on river speed, can pump water through a lift of 5 metres at a maximum rate of some 24 cubic metres per hour; and

(ii)   the smaller 'Low Cost' version (swept area up to 3.75 square metres, shown in figure 2.6), which, depending on river speed, can pump water through a lift of 5 metres at a maximum rate of about 6 cubic metres per hour.

Both machines can pump through higher lifts at lower delivery rates. As a rough indication, the Mark 1 turbine operating for eight to ten hours a day is capable of irrigating an area of 3 hectares and the low cost version, operating for the same period, can irrigate a plot of 1/2 to 3/4 of a hectare. The required sizes of machine to pump a specified water output of 3.6 cubic metres per hour are shown, for varying conditions of current speed and height of lift, in Figure 1.1.

As a very approximate guide, the costs of manufacturing the two versions in Southern Sudan (excluding delivery pipe and installation) at 1982 prices are:

Mark 1 version   :   US$ 5,000
Low Cost Version :   US$ 2,000

1.3 Alternative Water-Lifting Methods

The first step to determining whether the purchase of a WCT is likely to be good value for money is to establish whether any type of water pumping for irrigation is likely to be economically cost-effective.

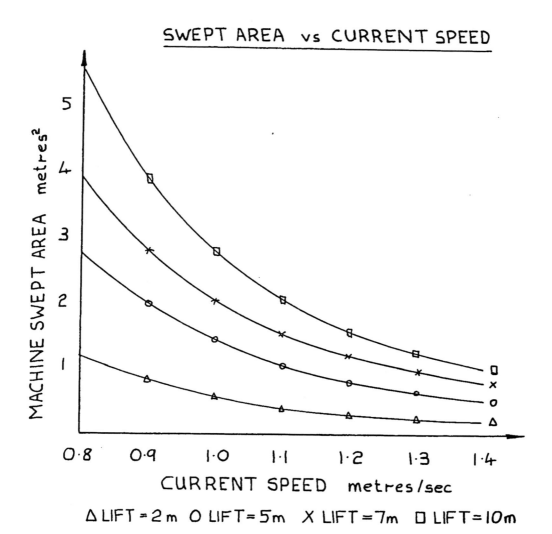

Note that the largest machine so far tested has a swept area of 5 square metres.

Assumptions:

(i)   Machine size is that required to achieve water output at end of delivery pipe of 1 litre/sec (3.6 cubic metres/hour).

(ii)   Overall system efficiency 7 per cent.

FIGURE 1.1:   Required machine size to pump (3.6 cubic metres per hour) as a function of water current speed and height of lift.

This question is discussed at length in Section 3.2. Assuming the answer is positive, two further questions immediately follow:

(i) if pumping for irrigation already occurs, can the bad features of existing water lifting devices be improved - or should consideration be given to introducing a new technology to the area?

(ii) alternatively, if there is no irrigation pumping at present, what is the most appropriate technology to use

These are the first questions which must be asked. The handbook can be used properly once the shortcomings of existing methods have been identified, a decision has been taken to consider introducing a new technology and water current turbines appear to be technically viable. Ideally , a range of other options should then be considered. The alternatives (classified according to power source) may include:

(i)   technologies based on human or animal power:
       (a) traditional water lifting devices such as
            - waterwheel
            - shadouf
            - archimedean screw
       (b) handpumps

(ii)   technologies based on renewable energy sources:
       (a)   water current turbines (WCTs)
       (b)   wind pumps
       (c)   solar pumps

(iii) technologies based on fossil fuels:
       (a)   diesel pumps

The treatment in this handbook of these alternatives is somewhat uneven. This lack of balance is most seriously evidenced by the inadequate account which is taken of traditional water-lifting devices. This deficiency reflects the lack of detailed technical and economic information on these traditional methods. (See Kennedy and Rogers, 1985 reference 5, p113) for a compendium of the information which is available). Fortunately, detailed information on the relative cost-effectiveness of human, animal, wind, solar and diesel power for water lifting has become recently available (See references 3 and 4, p113) and this together with the specific evidence from Southern Sudan (See references 1 and 2, p113) which compares water current turbines to small diesel pumps is the main data source.

4

1.4   Background

ITDG started working on the extraction of energy from river currents in 1978, when work at Reading University Applied Research Section demonstrated that a vertical axis 'Darrieus' type rotor would operate efficiently in water. Early testing of model rotors on the River Thames was funded by a grant from the Hilden Trust and continued throughout 1979.

In April 1980, the Royal Netherlands Government provided a grant for an extended field test of a water pumping turbine with a rotor shaft power of 1 kW. ITDG's prototype water pumping turbine was launched onto the White Nile at Juba in November 1980 and operated more than 7,000 hours over three dry seasons, irrigating a commercial horticultural garden for the last two seasons.

Further funding from the Royal Netherlands Government has enabled the construction and testing of further turbines with a variety of rotor and pump designs.

As a result of this work, the two sizes of turbine described previously (and in more detail in Chapter 2) have been developed and manufacturing drawings prepared by ITDG.

1.5   The Structure of this Handbook

The order of the contents of this handbook and the emphasis given to each section reflect the central objective of producing a document which identifies a method for technical, economic and social appraisal which can be put to general use. To illustrate how this method may be applied, and why, in some circumstances, certain factors may be critical, frequent use is made of examples - particularly from the Southern Sudan case study. Details of the Sudan example, and of the more complex aspects of the technical design considerations and method of economic appraisal, are included as appendices to the main text.

CHAPTER TWO

Technical Aspects

2.1 River Currents as an Energy Source
2.1.1 Introduction

It is important to be clear from the start that this book is
concerned with the extraction of kinetic energy from a freely
flowing river or canal in situations where it is impractical
(on engineering or economic grounds) to create a static head
of water by the construction of any sort of dam or barrage.
Figure 2.1 shows the geographical situation we are concerned
with. The river shown is up to 400 metres wide and flows
between low banks on an almost flat plain. Compared to the
energy available from a static head of water, river currents
are a very diffuse energy source. For example, a river speed
of one metre per second is equivalent, in energy terms, to a
static head of only 50 mm. Thus any static head of water
available should always be exploited (using the relevant
technology) in preference to a freely flowing, river or
canal. Having said this, river currents have many advantages
as an energy source. They can provide a reliable and
predictable energy supply which is available 24 hours per
day. Relatively simple technologies can convert river
current energy to provide pumped water in sufficient
quantities for economically viable small-scale irrigated
agriculture.

2.1.2 Calculation of Power Available in Flowing Water

The energy flux (or power available) in flowing water can be
calculated from the following equation:

$$P_a = 1/2 \rho A V^3 \qquad \ldots\ldots[1]$$

P_a is power available   (Watts)

ρ is the density of water   (1000 kg/m^3)

A   is the area of flow perpendicular to
the current direction from which
power is to be extracted   (m^2)

V   is the water velocity   (m/s)

In practice, it is not possible to extract all the power
available in a river current for two reasons. First, to give
up all its kinetic energy the water would have to stop, which
clearly it cannot do in free stream. Second, some type of
turbine rotor (see Figures 2.6, 2.7 and 2.8) must be used to
convert the water's kinetic energy into shaft power, and this
rotor is bound to be subject to drag forces which will
dissipate some of the power. Adding in a constant to
represent the conversion efficiency from energy flux in the
flowing water to power output of the turbine shaft, our
equation becomes:

FIGURE 2.1: The White Nile at Rejaf Near Juba.

Rotor Type	Economy of materials, including supporting frame	Speed of output	Ease of construction	Suitability for shallow rivers	Position of power take-off for driving c/f pump	Position of bearings	Ability to cope with debris	Comments
Floating waterwheel	1	1	9	9	8	10	1	Large quantity of materials used for comparable power output, not tested by ITDG (See references 6,9,19).
Vertical axis Darrieus	5	5	2	8	10	6	6	More suitable for larger machines of over 1 kW shaft power.
Horizontal axis Darrieus	4	10	3	8	0	0	8	Bearing and power take-off problems stopped development of this rotor.
Inclined axis propeller	7	7	6	6	9	6	6	At present this rotor is best choice for machines less than 1 kW shaft power.
Horizontal axis propeller	7	9	6	4	0	0	8	Not tested by ITDG due to anticipation of bearing problems from horizontal axis Darrieus tests.
Trailing propeller rotor	8	8	6	5	8	6	7	May be best choice once testing of machine as shown in Figure 2.9 is completed.
Half submerged propeller	8	6	6	8	8	8	1	Lower coefficient of performance due to splashing.

TABLE 2.1: Comparison of Alternative Turbine Rotors.

$$P_s = 1/2 \, \rho \, A_s \, V_s^3 \, C_p \qquad \ldots\ldots\ldots [2]$$

P_s is the turbine shaft power   (Watts)

A_s is the area of water current (perpendicular to the current direction) interrupted by the turbine rotor, known as the swept area (m^2).

V_s is the free stream velocity measured at least two rotor diameters upstream from the turbine   (m/s)

C_p is the coefficient of performance of the turbine rotor

From this equation it can be seen that there are three factors which affect the shaft power output of the turbine:

1. The turbine shaft power is proportional to the cube of the upstream current velocity. This means that, if the water speed is doubled, the rotor power output will be increased by a factor of eight. Figure 2.2 shows how the power output of a rotor of 3.75 m^2 swept area and coefficient of performance 0.25 would vary with the current speed. Note the very low output at current speeds less than 1 m/s.

2. The turbine shaft power is directly proportional to the rotor swept area. Thus a turbine of swept area 1.9m^2 would have a power output of half that of the machine in Figure 2.2. The largest swept area of any machine so far tested is 5m^2 which would produce 625 Watts in a current speed of 1 m/s and 1 kW at 1.17 m/s assuming a C_p value of 0.25.

3. The power output is also directly proportional to the coefficient of performance. As already mentioned, it is impossible to extract all the energy from the flowing water because the water which has passed through the rotor must move away from it and therefore must still have some kinetic energy. It can be shown theoretically (See reference 6, p113) that the maximum coefficient of performance is 0.59 for a machine operating on lift forces such as a propeller or Darrieus rotor and 0.33 for a machine operating on drag forces such as a floating undershot water-wheel in free stream. Our testing of Darrieus and propeller type rotors has indicated that <u>under typical field manufacture and use conditions</u> their coefficient of performance will be between 0.2 and 0.25, depending on the river speed and manufacturing quality achieved (See References 6,7 and 8, p113).

From the above we can see that to obtain the maximum shaft power output we should use the most efficient type of rotor available, make it sweep as large a cross sectional area of water current as possible and, most importantly, place it in the fastest current speed which can be found.

2.1.3 Minimum Useful Current Speed

To extract a given amount of power the machine becomes larger as the current speed decreases. A machine in a current speed of 0.5 m/s would have to be eight times the size of one in a current speed of 1 m/s to produce the same shaft power (see Figure 2.2.).

As can be seen from Figure 2.2, the level of energy flux in river currents of less than 0.8 m/s is so low that there would have to be very special economic conditions to justify the construction of a machine large enough to extract useful amounts of power.

The possibility of using a duct to artificially increase the water velocity through the turbine rotor has been investigated and found to produce a small improvement in energy extracted per unit area of current intercepted. However the considerable increase in capital cost and the increased difficulties of transporting and manoeuvering the machine eliminate the ducted free stream turbine from further consideration as a low cost water pumping turbine.

2.1.4 Minimum Useful Depth

Having established the minimum useful current speed from the point of view of energy extraction, we now turn our attention to the depth of water required. To do this it is necessary to start at the final use and determine the quantity of pumped water required. Once the required water output, the total pumping head and the current speed have been determined (by methods explained in later chapters) the required turbine swept area can be found by working back through the various components of the machine. The estimation of the machine's overall system efficiency is dealt with in Section A1.4. but at this stage it can be said that the ITDG 'Low Cost' water current turbine (see Figure 2.6) will convert up to 7 per cent of the energy flux through its rotor into hydraulic output at the end of the water delivery pipe.

The hydraulic output power of the system (P_o) is calculated from:

$$P_o = QH_s g \qquad \ldots\ldots[3]$$

Q	is the water delivery	(litres/second)
H_s	is the static pumping head or lift	(metres)
g	is the acceleration due to gravity	(9.81 m/s^2)

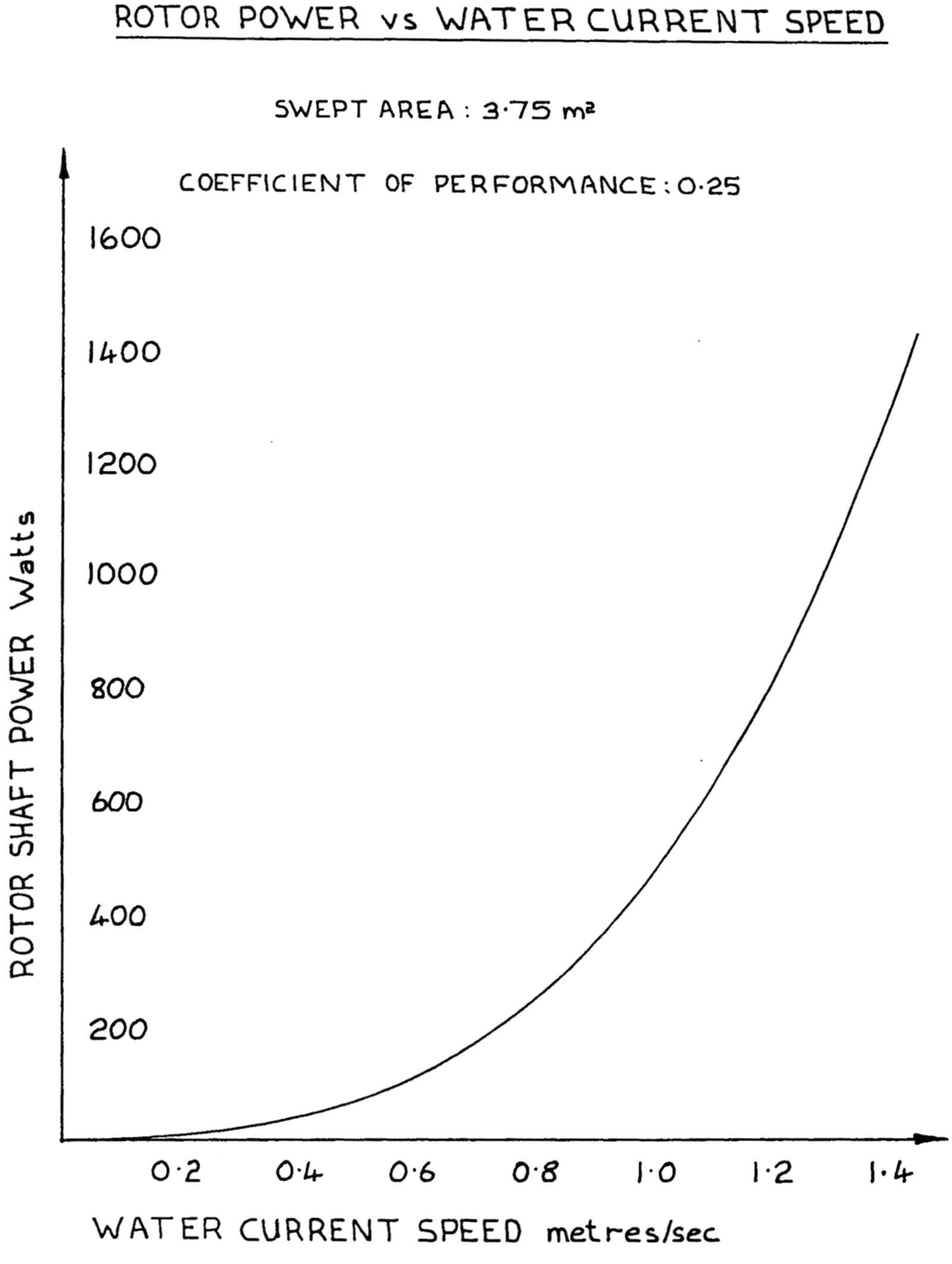

ROTOR POWER vs WATER CURRENT SPEED

SWEPT AREA : 3·75 m²

COEFFICIENT OF PERFORMANCE : 0·25

FIGURE 2.2:   Graph Showing Rotor Power as a Function of Water Current Speed.

Experience has shown that 1 litre/second is the minimum useful water output for a vegetable plot irrigated by an earth channel distribution system. Hence taking a static pumping head of, say, 7 metres we see that the minimum useful hydraulic output power would be 6.9 Watts. The <u>overall system efficiency</u> is the ratio between the system output and the power available in the water flowing through the turbine rotor:

$$\eta = \frac{P_o}{P_a}$$

ie $\quad \frac{1}{2}\rho A_S V^3 = \frac{QH_Sg}{\eta} \quad \quad \quad ......[4]$ from [1] and [3]

Hence, in a current speed of 0.8 m/s, the swept area required is 3.85 square metres (see Figure 1.1). By similar calculation for different lifts and water current speeds, curves such as Figure 1.1 can be produced. To instal a 'Mark 1' or 'Low Cost' turbine of this swept area a river depth of at least 2.7 metres is required if the water speed is only 0.8 m/s. The required water depth will be less in faster current speeds due to the reduced swept area (ie a smaller rotor) fitted to the machine. In a current speed of 1.8 m/s a water depth of 1.6 metres is required to extract 80 watts hydraulic output from the machine.

From the above we can conclude that a water current must have a velocity of at least 0.8 m/s and a depth of at least 1.8 metres before useful quantities of power can be extracted by turbines operating in free stream. If the machine is to pump water, these river conditions must exist, within 25 metres from the river bank. If the water current speed is greater than 1.8 m/s then the two designs discussed here would require some detail design modifications such as float size, mooring arrangement, rotor diameter and transmission ratio.

To put this into context: One the White Nile near Juba current speeds vary between 0.75 m/s and 1.5 m/s, depending on the site and season. In most places the river is at least 3 metres deep within 10 metres of the bank and the change in level is only about one metre over the year. The variation in current speed at a given site over the dry season is normally less than 15 per cent.

2.2 Site Selection

In the last section we established the minimum river speed and depth for any form of kinetic energy extraction to be viable. Like conventional water powered devices, river current turbines are a site-specific technology. For example, the type of pump fitted to the machine will depend on the total delivery head, and, as already seen, the diameter of the machine rotor will depend on the river current speed.

Before starting work on the construction of a turbine, it is necessary to survey the proposed site for the machine to provide the following basic information:

(i)   the quantity of water required and hence the delivery from the machine in litres per second;

(ii)   the maximum static pumping head or lift required from the river surface to the delivery pipe outlet;

(iii)   the diameter and length of the delivery pipeline from the machine to the outlet at the field;

(iv)   the maximum and minimum river current speed over the months that the machine will be used;

(v)   the minimum river depth at the position where the turbine will operate and the minimum depth at the river bank;

(vi)   enviromental hazards such as floating debris, river traffic, etc.

The delivery of water required will depend on the following factors:

(i)   the areas to be irrigated;

(ii)   the water requirements of the crop being grown;

(iii)   the local climate;

(iv)   the type of distribution system used, eg earth channels, hose pipes, etc;

(v)   whether a water storage tank is available;

(vi)   the number of hours the machine will be run for each day.

For example, at a 1/4 hectare garden growing vegetable (salads, ocra etc) at Juba in Southern Sudan with no water storage and an earth channel distribution system, a delivery of 1 litre/second was necessary to water the garden in six or seven hours per day.

The pipeline details are important because friction in the pipes produces an additional resistance for the pump to overcome and this resistance must be added to the static head or lift to obtain the total or dynamic head which the pump must generate. Large friction losses due to too narrow or too long a delivery pipeline can reduce the system's efficiency considerably, resulting in an increase in the size and cost of the turbine.

2.3 Measurement of River Current Speed

The river current speed can vary by as much as 10 per cent
within 30 or 40 metres up or down stream from a given spot.
Bearing in mind that a 10 per cent increase in river speed
gives a 30 per cent increase in rotor shaft power (see
equation 1, p6) the importance of accurate current speed
measurement for selecting the best site will clearly be
appreciated.

Accurate speed measurement is also necessary to select
the correct rotor swept area to ensure that the required
amount of power (and not too much as this might damage the
transmission) is produced.

For anyone involved in serious testing or production of
turbines, a propeller meter with an audible counter such as
the Braystoke BFM001 is the ideal instrument. This type of
instrument not only gives an average river speed (over a
variety of timing periods) which is accurate to plus or minus
one per cent, but also gives an idea of the steadiness of the
current by means of its audible counter. The meter should be
suspended at the proposed turbine site at the position of the
rotor centre.

If this type of current meter is unavailable, the river
speed should be measured by throwing in a piece of wood and
timing it to travel between two pairs of posts placed at
least 50 metres apart on the river bank, (each pair of posts
are arranged to give a line of sight at right angles to the
current direction). Another useful method, providing the
river does not run due east-west, is to use the sun's
reflection in the water (in the morning if you are on the
west bank and the evening if you are on the east bank) as the
timing mark. Simply throw in the wood upstream from where
you are standing and start a stopwatch (or note the time on a
watch with a second hand) when the wood crosses the sun's
reflection. Move quickly to a spot a measured distance of 50
metres downstream and stop the watch when the wood crosses
the sun's reflection as seen from your new position. Do the
speed tests on a day when there is little wind. In view of
Figure 2.2 the importance of accurate speed measurement
cannot be overstressed.

2.4 Water Pumping Turbine System Design

So far, the only consideration we have given to the design of
a water pumping turbine is to decide on the required rotor
swept area. The rotor is only one element of a machine which
delivers water to the river bank. Figure 2.3 shows all the
elements of the machine's design. Each of these has to be
considered in turn, and their detailed design will vary
depending on the site conditions and the materials, parts and
production processes available locally. In the following
sections the function of each of the elements of the
machine's design is explained and alternative designs
and materials are discussed.

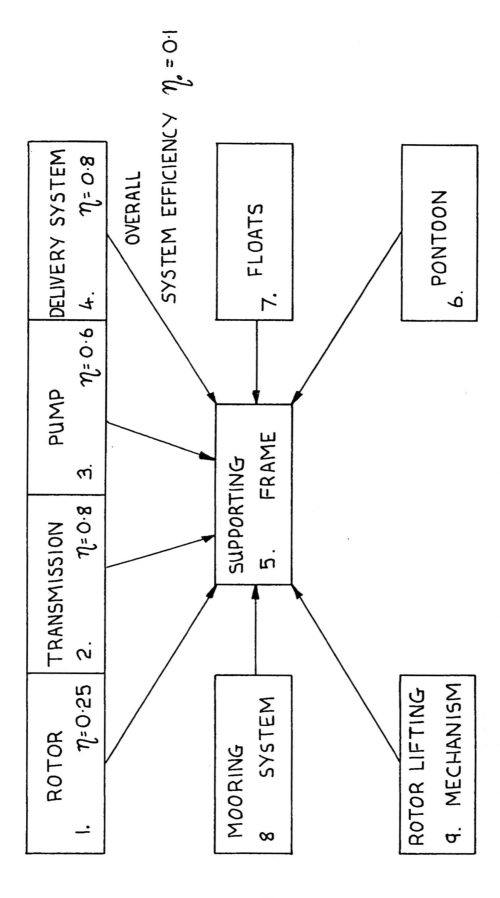

FIGURE 2.3: Main Components of Water Pumping Turbine.

15

The elements of the design which are directly concerned with converting the water current energy into pumped water are the rotor, transmission, pump and delivery system. At all stages in this conversion there are losses, and much of the design effort has been to reduce these losses to the minimum consistent with reasonable capital cost. The less efficient the various components, the lower the overall efficiency and hence the larger (and more expensive) the machine required to pump a given amount of water. The overall system efficiency is the product of the efficiencies of the system components and is equal to the hydraulic power output divided by the power available (see 2.1.4).

ie

$$\frac{QH_s g}{1/2 \rho A_s V^3} = \eta_{ROTOR} \times \eta_{TRANS.} \times \eta_{PUMP} \times \eta_{DELIVERY\ SYSTEM}$$

Figure 2.3 gives typical efficiencies for the components of the 'Mark 1' machine, resulting in an overall system efficiency of about 0.1 (or 10 per cent). Due to its lower pump efficiency the 'low cost' machine has a system efficiency of about 0.07 (7 per cent).

The 'Mark 1' (Figure 2.4) and 'low cost' (Figure 2.6) water pumping turbines developed by ITDG at Juba Boatyard in Sudan have already been briefly described in Section 1.2. The 'low cost' machine was designed around materials and parts then available in southern Sudan, and its capital cost is kept to a minimum in the hope of making it affordable by smallholders. The relatively high output 'Mark 1' machine uses a pump and transmission which is specially imported into southern Sudan, but which would be locally available in many countries. Ferrocement floats are used on this machine as a more durable alternative to oil drums.

These two designs represent the state of the art in terms of ITDG's work. All the elements of each of the designs have undergone sufficient field testing for the machines to be constructed with confidence for extended field testing (and subsequent modification to suit local requirements, site conditions, construction materials, production processes, etc) before pilot commercial manufacture. It must be appreciated that water current turbine technology is in its infancy and the development of the new ideas discussed in subsequent sections is expected to result in very substantial cost reductions and improvements in operation.

Manufacturing drawings of both the turbine designs discussed here have been completed and are available with technical assistance from ITDG by mutual agreement. Readers considering experimenting with this technology should also study the work of the Danish and Sudanese Guide and Scout associations on hydrostatic coil pumps (see reference 9, p113).

Before commercial manufacture is contemplated, an absolute minimum of two complete dry seasons testing of, say, four or five machines, with local farmers, is recommended. It should be noted that the highest current speed in which these machines have been tested is 1.4 m/s. In higher current speeds various parts of the design may need to be strengthened and a larger pontoon may be necessary. The first machine should be installed in a current speed less than 1.2 m/s, at least until operating experience of handling the mooring system, winches and delivery pipe has been gained.

2.5 Turbine Rotor

2.5.1 Choice of Turbine Rotor

As already mentioned, the function of the turbine rotor is to convert as much as possible of the kinetic energy flux through it into useable shaft power. The range of possible turbine rotors is similar to the different types used to extract energy from the wind. There are two basic types of rotor operating on different principles.

1. Machines which have their effective surfaces moving in the direction of the current and are pushed round by the drag of the water, eg undershot water wheel as shown in Figure 2.5.

2. Machines which have their effective surfaces moving at an angle to the direction of the water and operate on lift forces, eg propeller rotor and Darrieus rotor as shown in Figures 2.7 and 2.8 in various alternative arrangements.

Figures 2.5, 2.7 and 2.8 show the dimensions and depth of water required for each of the rotors to produce the power output shown in Figure 2.2. For comparison it is assumed that all the rotors shown have the coefficient of performance but of the designs tested by ITDG the propeller rotor was the most efficient. Reference 19, p113 compares the coefficients of performance of various rotors.

Table 2.1 shows the relevant criteria by which types of rotor might be selected. The possible range of rotors are rated from zero to 10 on each of the selected criteria. Zero represents an as yet unsolved problem which rules that particular type of rotor out for the time being. One indicates a particularly poor performance. A high rotational speed is desirable to minimize the cost and complexity of the transmission.

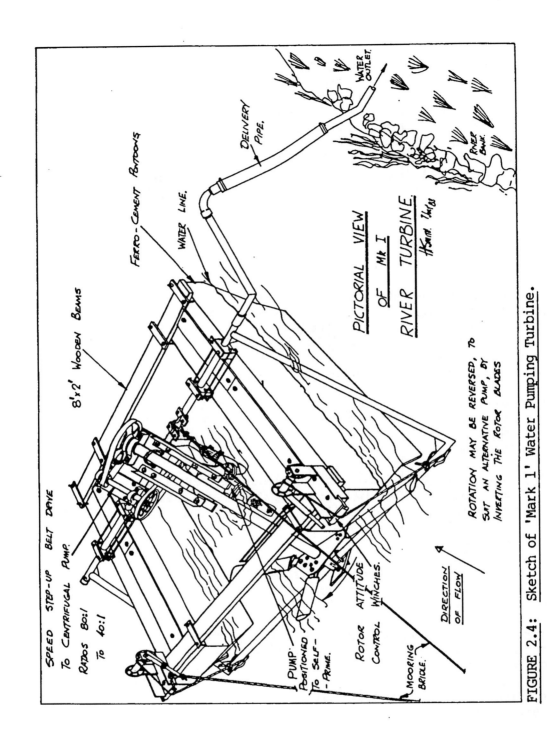

Speed step-up belt drive
To Centrifugal Pump.

Ratios 80:1
To 40:1

8'x2' Wooden Beams

Ferro - Cement Pontoons

Water Line.

Delivery Pipe.

Water Outlet.

River Bank.

PICTORIAL VIEW
OF Mk I
RIVER TURBINE.
Hkint. Unless

ROTATION MAY BE REVERSED, TO
SUIT AN ALTERNATIVE PUMP, BY
INVERTING THE ROTOR BLADES

Pump.
Positioned
To Self-
Prime.

Rotor Attitude
Control
Winches.

Mooring Bridle.

Direction
of Flow

FIGURE 2.4:   Sketch of 'Mark 1' Water Pumping Turbine.

18

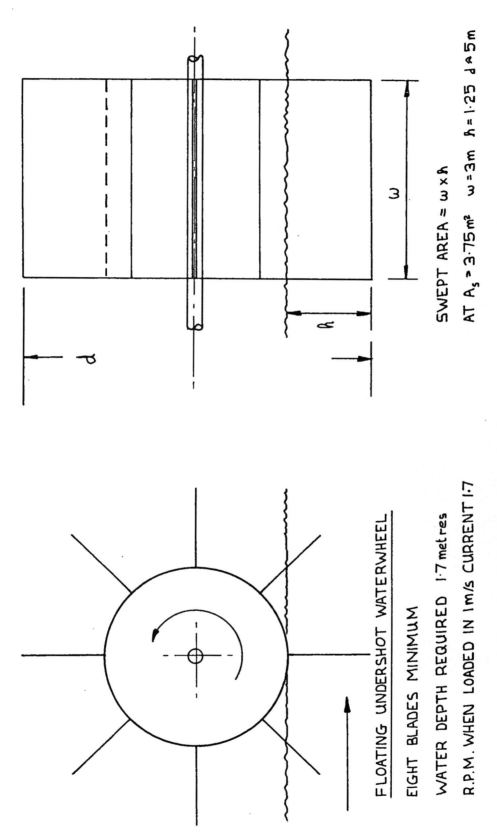

SWEPT AREA $= \omega \times h$

AT $A_s = 3.75m^2$ $\omega = 3m$ $h = 1.25m$ $d \approx 5m$

FLOATING UNDERSHOT WATERWHEEL

EIGHT BLADES MINIMUM

WATER DEPTH REQUIRED 1·7 metres

R.P.M. WHEN LOADED IN 1m/s CURRENT 1·7

FIGURE 2.5: Sketch of Floating Undershot Waterwheel

19

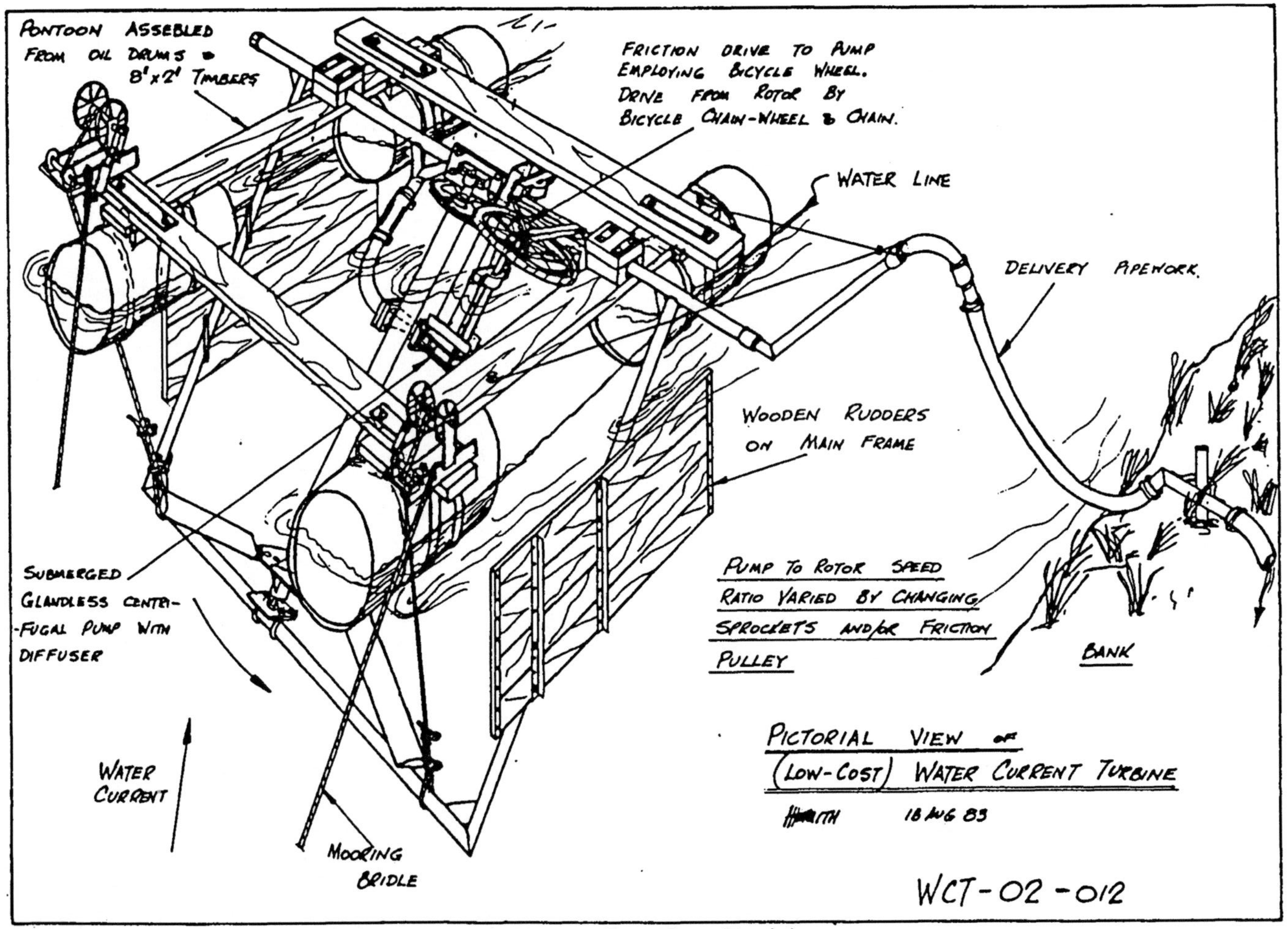

PONTOON ASSEBLED FROM OIL DRUMS & 8'x2' TIMBERS
FRICTION DRIVE TO PUMP EMPLOYING BICYCLE WHEEL. DRIVE FROM ROTOR BY BICYCLE CHAIN-WHEEL & CHAIN.
WATER LINE
DELIVERY PIPEWORK.
WOODEN RUDDERS ON MAIN FRAME
SUBMERGED GLANDLESS CENTRI-FUGAL PUMP WITH DIFFUSER
PUMP TO ROTOR SPEED RATIO VARIED BY CHANGING SPROCKETS AND/OR FRICTION PULLEY
BANK
WATER CURRENT
MOORING BRIDLE
PICTORIAL VIEW OF (LOW-COST) WATER CURRENT TURBINE
18 AUG 83
WCT-02-012

FIGURE 2.6: Sketch of 'Low Cost' Water Pumping Turbine

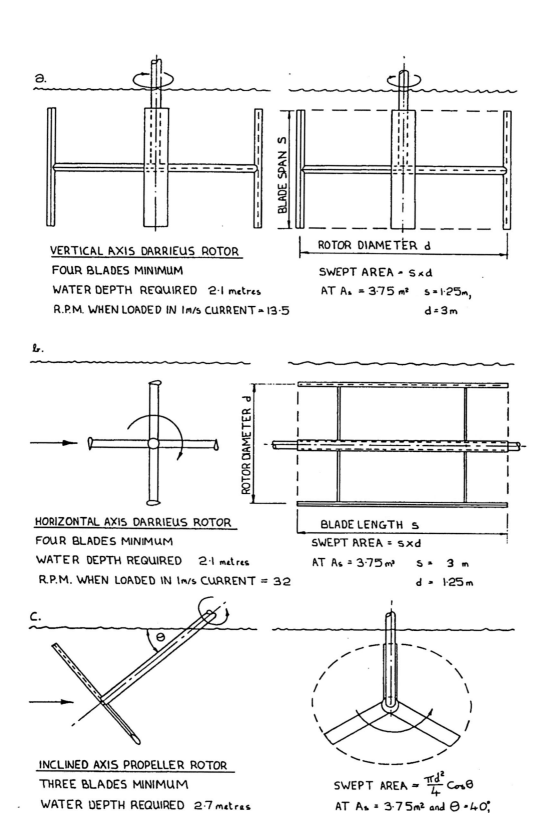

a.

BLADE SPAN s

ROTOR DIAMETER d

<u>VERTICAL AXIS DARRIEUS ROTOR</u>

FOUR BLADES MINIMUM

WATER DEPTH REQUIRED   2·1 metres

R.P.M. WHEN LOADED IN 1m/s CURRENT = 13·5

SWEPT AREA = s×d

AT A_s = 3·75 m²   s = 1·25m,

d = 3 m

b.

ROTOR DIAMETER d

<u>HORIZONTAL AXIS DARRIEUS ROTOR</u>

FOUR BLADES MINIMUM

WATER DEPTH REQUIRED   2·1 metres

R.P.M. WHEN LOADED IN 1m/s CURRENT = 32

BLADE LENGTH s

SWEPT AREA = s×d

AT A_s = 3·75 m²   s = 3 m

d = 1·25 m

c.

θ

<u>INCLINED AXIS PROPELLER ROTOR</u>

THREE BLADES MINIMUM

WATER DEPTH REQUIRED 2·7 metres

R.P.M. WHEN LOADED IN 1m/s CURRENT = 22·9

SWEPT AREA = $\frac{\pi d^2}{4} \cos\theta$

AT A_s = 3·75 m² and θ = 40°,

d = 2·5 m

<u>FIGURE 2.7:</u>   <u>Sketches of Alternative Turbine Rotors.</u>

21

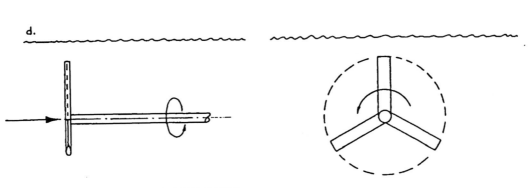

HORIZONTAL AXIS PROPELLER ROTOR

THREE BLADES MINIMUM

WATER DEPTH REQUIRED = 3 metres

R.P.M. WHEN LOADED IN 1m/s CURRENT = 26.2

SWEPT AREA = $\frac{\pi d^2}{4}$

AT As = 3.75 m²   d = 2.19 m.

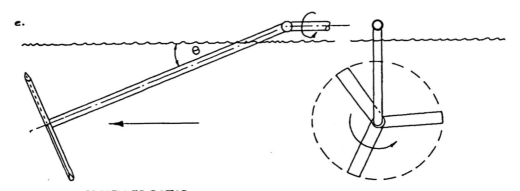

TRAILING PROPELLER ROTOR

THREE BLADES MINIMUM

WATER DEPTH REQUIRED ≈ 2.9 metres

R.P.M. WHEN LOADED IN 1m/s CURRENT = 25.5

SWEPT AREA = $\frac{\pi d^2}{4} \cos\theta$

AT As = 3.75 m² and θ = 20°, d = 2.25 m.

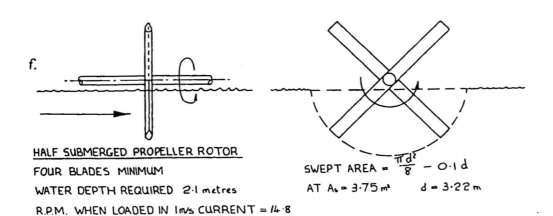

HALF SUBMERGED PROPELLER ROTOR

FOUR BLADES MINIMUM

WATER DEPTH REQUIRED 2.1 metres

R.P.M. WHEN LOADED IN 1m/s CURRENT = 14.8

SWEPT AREA = $\frac{\pi d^2}{8} - 0.1d$

AT As = 3.75 m²   d = 3.22 m

FIGURE 2.8:   Sketches of Alternative Turbine Rotors.

From Table 2.1[*] it can be seen that for machines under 1 kW shaft power, the choice is between the inclined axis propeller rotor (Figure 2.7c) and the trailing propeller rotor (Figure 2.8e). These are the only rotors with consistently high ratings in all categories. Four machines have been successfully tested with inclined axis propeller rotors, and between them have run for more than 4,000 hours. A trailing rotor has been built but due to lack of time and funds was only tested for 250 hours and is therefore not proven yet.

A water pumping turbine using a trailing rotor is expected to have a lower materials cost than the 'Mark 1' type design (see Figure 2.4), especially on sites where the turbine can be positioned within 15 metres of the bank. The suggested arrangement of this machine is shown in Figure 2.9. It should also be possible to maintain this machine from the river bank.

However, until further testing of the trailing rotor design is carried out, the inclined axis propeller turbine is the most suitable tried and tested rotor design for machines under 1 kW shaft power.

A more detailed comparison of the performance of different rotors is given in Appendix 1.1.

2.5.2 Rotor Construction Materials

Much time and effort has been spent in investigating different materials for rotor construction. During this time the following materials have been tried:

(i)   solid aluminium alloy;
(ii)   laminated hardwood sheathed with glass fibre reinforced plastic (GRP);
(iii)   steel spar with polyurethane foam filled GRP fairing;
(iv)   untreated hardwood;
(v)   ferrocement (a) untreated, (b) painted, (c) sheathed with Al alloy sheet;
(vi)   steel spar, timber fairing sheathed with Al alloy sheet.

Of these alternatives, all have proved structurally satisfactory except untreated hardwood, which warped and cracked in the water. From the performance point of view surface finish is critical, and any deterioration causes drastic shaft power reduction. This is because the blade velocity of lift-powered rotors is twice (in the case of the Darrieus rotors) or three times (in the case of propeller rotors) that of the river current, and so drag produced by

* See page 8

23

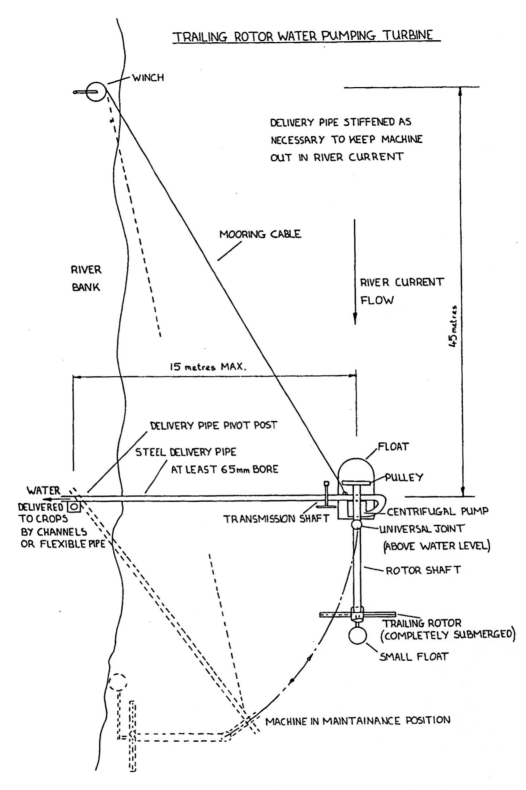

TRAILING ROTOR WATER PUMPING TURBINE

WINCH

DELIVERY PIPE STIFFENED AS
NECESSARY TO KEEP MACHINE
OUT IN RIVER CURRENT

MOORING CABLE

RIVER
BANK

RIVER CURRENT
FLOW

45 metres

15 metres MAX.

DELIVERY PIPE PIVOT POST

STEEL DELIVERY PIPE
AT LEAST 65mm BORE

FLOAT

PULLEY

WATER
DELIVERED
TO CROPS
BY CHANNELS
OR FLEXIBLE PIPE

TRANSMISSION SHAFT

CENTRIFUGAL PUMP

UNIVERSAL JOINT
(ABOVE WATER LEVEL)

ROTOR SHAFT

TRAILING ROTOR
(COMPLETELY SUBMERGED)

SMALL FLOAT

MACHINE IN MAINTAINANCE POSITION

FIGURE 2.9: Sketch of Trailing Rotor Water Pumping Turbine.

24

surface friction is a very important consideration. The only materials which maintained their surface finish and high level of performance were GRP and Al alloy. Some of the GRP sheathed blades on the original prototype machine were still in good condition after 7,000 hours of running and nearly three years in the water. Unfortunately, the polyester resins and catalyst required for GRP blade manufacture are difficult and potentially hazardous materials to transport, store and use in tropical conditions and may not be widely available. The Al alloy sheathed blades have been tested on four different machines but none of them has yet run for more than 2,000 hours. If suitable sheet material is not available (19 swg or 1 mm thick), then body panels from discarded Land Rovers can be used. Note, however, that it is essential to remove all traces of paint and primer from the metal, as contact between the primer and water may cause serious surface pitting.

Various epoxy coatings are now available in Europe for ferrocement; these may be an alternative surface material to GRP or Al alloy but so far none has been tested on turbine blades. Use of these coatings would enable the development of twisted blades for the propeller rotors. Slightly twisted blades would improve the propeller rotor's self starting ability, and at least in theory should improve its performance. These gains are not considered likely to be large enough to make up for the increased difficulty of manufacturing a twisted version of the present Al alloy sheathed blades.

2.5.3 Rotor Bearings

The rotor shaft must be carried in bearings which support it in the correct position relative to the river current and allow it to rotate as freely as possible. If the shaft is to be supported at each end by a bearing mounted on a frame (see Figures 2.4 and 2.6), at least one of the bearings must allow some axial movement to take up flexing of the frame, and both must allow some misalignment to compensate for assembly errors or adjustment of the first transmission stage.

It is these requirements which have so far not been satisfied for rotors with both ends of the shaft under water and which have thus halted development of the horizontal axis Darrieus and horizontal propeller rotors.

The inclined axis propeller rotor has one bearing above the water for which a single row ball bearing is suitable. The bearing used is the grease-lubricated self-aligning type mounted in a cast iron pillow block as commonly used in agricultural equipment. This bearing provides axial location for the rotor shaft, takes the axial thrust on the rotor and takes the radial load due to the belt tension in the first stage of the transmission.

The bearing at the bottom end of the rotor shaft is underwater and hence must be water-lubricated. This bearing locates the hub end of the rotor shaft, takes a small radial load and allows some axial movement of the shaft relative to the frame. After experiments with timber and tufnol (a phenolic resin impregnated paper widely used in marine sterngear) running on steel or stainless steel, it became clear than any type of bearing with one rubbing surface harder than the other was impractical. The reason for this is that silt from the water becomes embedded in the softer of the two materials which then abrades the hard surface very quickly. Any water current which flows fast enough to drive a turbine is almost certain to be carrying quantities of silt similar to the Nile and so this problem is likely to be encountered everywhere to a greater or lesser degree. An acceptable solution to the bottom bearing has been found and is simply a steel pin mounted on the frame around which a mild steel insert in the end of the shaft rotates. The pin is easily made by cutting the head off a high tensile or stainless steel bolt. This bearing has proved to be satisfactory and the pin and insert will last at least 5,000 hours before requiring replacement. Appendix 1.2, on p88, outlines the expected bearing loads.

2.6 Transmission

The fraction of the river current energy extracted by the turbine rotor is available from the rotating turbine shaft which can exert a torque (or turning force) against a load. To drive a centrifugal pump (see Section 2.7) it is necessary to increase this speed of rotation, usually by a factor of between 50 and 100. Using modern flat belts it is possible to achieve this ratio in two stages with an intermediate shaft between the rotor and pump (see Figures 2.4 and 2.10). 'Poly V' belts were selected for the 'Mark 1' machine because of the high speed ratios obtainable (eg 10:1 per stage), and because they can run with the shafts at any angle without the need for idlers or crowned pulleys. It is also possible to manufacture the pulleys on an ordinary centre lathe (this is not the case, for example, with toothed belts). 'Poly V' belt transmissions have performed very well in the field tests. The belts are hard wearing, not badly affected by sun and rain and reasonably tolerant of misalignment. 'Poly V' belt transmission efficiency is about 90 per cent per stage, given reasonable shaft alignment. For rotor shaft powers up to 1 kW it is possible to use 'J' section belts (the smallest section size) on both stages, but to go up to, say 1,200 Watts it would be necessary to use the 'L' section belt on the first stage resulting in a considerable increase in cost. The method of calculating the required transmission ratio for a given site is explained in Appendix A1.3. on p89.

The only disadvantages of this type of transmission are,first, that the belts are likely to have to be specially imported into most countries and, second, that the cost of the belts, pulleys, and intermediate shaft and its bearings and adjusters is likely to amount to one third of the materials cost of the whole machine. For a small machine, such as the 'Low Cost' design, the cost of this transmission would be an even larger proportion of the total.

For this reason a transmission using cycle components, which should be locally available nearly everywhere, was designed and successfully tested for 2,800 hours. Figure 2.6 shows the 'Low Cost' transmission design and Figure 2.11 shows the experimental cycle component transmission from which it is designed. An overall speed ratio of up to 76:1 can be achieved in two stages. A 48-tooth front cycle sprocket mounted on the turbine rotor shaft drives, via a 1/4" wide bicycle chain, a 12-tooth sprocket fitted on a 28" rear bicycle wheel running in its own bearings. The inflated bicycle tyre then friction-drives onto the pump shaft. Good torque transmission is achieved onto a smooth turned pump shaft down to about 35 mm diameter. A knurled surface on the shaft simply produces very rapid tyre wear. On a smooth shaft of 50 mm diameter, the life of the bicycle tyre is about 750 hours, which represents one month's continuous running or three months at the usual watering rate of eight hours per day. This type of transmission is not practical at shaft power outputs of over 350 watts, due to stretching of the chain and accelerated wear on the small sprocket and cycle tyre, and it is this which limits the output of the 'Low Cost' machine. There is, however, no reason why this transmission should not be doubled with two bicycle chains and two tyres driving onto the same pump shaft. A separate spring-loaded tensioner (as fitted in Figure 2.11) would then be necessary on each chain.

Because the axes of rotation of the chain sprockets are not horizontal, two guides are necessary to stop the chain falling off. One is on the slack side of the chain to position it correctly just before it meshes with the small sprocket. The other is on the tight side where the chain meshes with the large sprocket. This last guide is only touched by the chain if it momentarily loses tension during starting or stopping.

On the 'Low Cost' design the chain is tensioned by simply moving the main shaft top bearing in its slots, but if this proves unsatisfactory a tensioner (using a complete freewheel assembly) as shown in Figure 2.11 can be added. An alternative arrangement of this transmission, using a leather-faced belt instead of the friction drive on the second stage, is described in reference 1 on p112, as are transmissions for reciprocating pumps.

FIGURE 2.10: 'Mark 1' Transmission and Pump.

FIGURE 2.11: 'Low Cost' Cycle Component Transmission.

2.7 Choice of Pump

2.7.1 Pump Types Available

There are basically two types of pump: rotodynamic such as the centrifugal pump which increases the pressure of the fluid being pumped by accelerating it in a confined space, and positive displacement such as the piston pump which entraps a volume of liquid and forces it through the delivery system by reducing the volume of the container (see Figure 2.12).

The pumps used on these machines are of the centrifugal type: machines with positive displacement pumps have been tested with limited success.

Centrifugal pumps have the following advantages:
1. A centrifugal pump gives a much better match with the turbine rotor than a piston pump. Figure 2.2 shows how the power output of a typical rotor varies with current speed, assuming a constant Cp of 0.25. Figure 2.13 shows how the rotor output varies as its rotational speed increases with increasing river speed. This assumes that the rotor always runs at its most efficient speed relative to the water current. Figure 2.13 also shows the power input requirements of a centrifugal pump and a positive displacement pump superimposed on the rotor output curve. Both pumps shown are chosen to absorb slightly less power than produced by the rotor at 0.8 m/s current speed.

As shown by the line on Figure 2.13 the power input to a positive displacement pump of given size varies in direct proportion to the number of strokes per minute. Thus, if a good match with the turbine rotor is achieved at the bottom end of the speed range, about half of the turbine's output power is wasted at the top design river speed. Similarly, if the pump was matched at 1.3 m/s it would stop the turbine at any river speed below this because the pump would require more power than the turbine produced. The match is not quite as bad as is indicated at first glance in Figure 2.13. Because only half the rotor's power was being abosrbed at 1.3 m/s, the rotor (and hence pump) speed would increase relative to the current and less power would be produced until a balance was reached. Thus, at 1.3 m/s river speed the reciprocating pump would actually be operating at about 34 strokes per minute rather than at 30 as indicated.

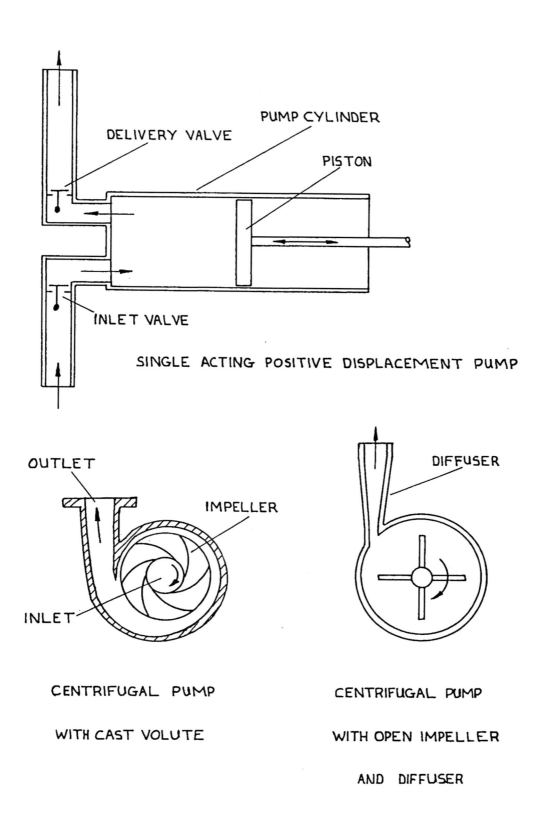

DELIVERY VALVE

PUMP CYLINDER

PISTON

INLET VALVE

SINGLE ACTING POSITIVE DISPLACEMENT PUMP

OUTLET

IMPELLER

INLET

DIFFUSER

CENTRIFUGAL PUMP

WITH CAST VOLUTE

CENTRIFUGAL PUMP

WITH OPEN IMPELLER

AND DIFFUSER

FIGURE 2.12: Positive Displacement and Centrifugal Pumps.

31

The power input requirement of a reciprocating pump can be changed by altering the stroke (and hence volume of water delivered) which must be adjusted on installation to get the best match possible at a given site. In practice, however, the current speed at any real site varies on a minute to minute cycle as well as on an annual one, and so in order to avoid endless stalling of the turbine the stroke must always be set short resulting in a poor system efficiency.

The power input to a centrifugal pump is approximately proportional to the cube of its rotational speed. Hence its characteristic is similar to the rotor output curve. The reason that the two lines are not parallel is that the efficiency of a centrifugal pump increases as its rotational speed increases. The pipe friction in the delivery system also affects the match (and this is discussed in detail later on), but in general it can be said that a centrifugal pump will provide a good match to the turbine over a wide range in river speed without any change in transmission ratio being necessary. Once the transmission ratio has been correctly set for a given rotor diameter and delivery head, the pump speed simply increases or decreases proportionally to the river speed allowing the turbine rotor to run near its most efficient speed relative to the current without any stalling problems.

2.   A centrifugal pump has a very low starting torque as, until it has reached a high enough speed for delivery to start, the only energy required is that to turn the pump shaft in its bearings and the impeller in the water. A positive displacement pump, on the other hand, requires a very high torque to start it, as its piston must be moved against the friction of the piston in the cylinder and the pressure of the column of water in the delivery pipe. In practice, this means that machines driving centrifugal pumps will self-start, whereas machines driving positive displacement pumps have to be started by hand after the delivery pipe has been drained.

3.   The centrifugal pump requires no valves, whereas a double acting piston pump (or two single acting piston pumps) requires four. These valves are not only expensive (even if locally made using pipe unions), but also require replacement and are an additional possible source of trouble.

4.   If the centrifugal pump is arranged with its impeller below water level it will self prime and can be run without a shaft seal.

5. The centrifugal pump has no rubbing parts in contact with the water if the shaft seal is replaced with a throttle bush. The rubbing surfaces in a piston pump wear quickly due to the sediment in the river water. In the case of closed impeller type pumps, the clearance between the impeller and case is relatively quite small. Eventually, continued running in water with course sediment will reduce the efficiency slightly due to this clearance increasing and allowing circulation in the pump. This is only likely to have a noticeable effect when pumping to a high head. An open impeller type pump has greater clearances but much lower efficiency.

6. The centrifugal pump delivers water in a steady flow, thus minimising friction effects in the delivery pipe. Reciprocating pumps deliver an unsteady flow which is continuously accelerated and slowed down in the pipe. To reduce these additional friction losses it is necessary to fit an air receiver to the pump outlet to smooth the flow in the delivery pipe.

There are only two disadvantages when using centrifugal pumps in this application:

1. A centrifugal pump must rotate very much faster than the turbine shaft. This type of pump generates head by acclerating the fluid from the centre to the outside of a rotating impeller inside a cylindrical or spiral casing. The pressure developed by the pump is proportional to the square of the peripheral f l u i d velocity at the outside of the impeller, and therefore to generate a given head the smaller the pump the faster it must turn. To generate a head of 7 metres or so a 150 mm diameter impeller must be rotated at about 1,300 rpm. Since the on-load turbine rotor speed is typically of the order of 20-40 rpm, the need for a transmission with a high speed ratio can be seen. As mentioned in Section 2.6, the disadvantage is the cost of the components necessary to assemble the transmission which can amount to one third of the materials used.

2. Centrifugal pumps are best suited to low head high delivery sites. All sites at which these turbines have been tested so far fall into this category, but there will be sites where the head required is greater than the maximum which the pump can generate (10 metres in the case of the 'Low Cost' machine or 25 metres in the case of the 'Mark 1' machine).

33

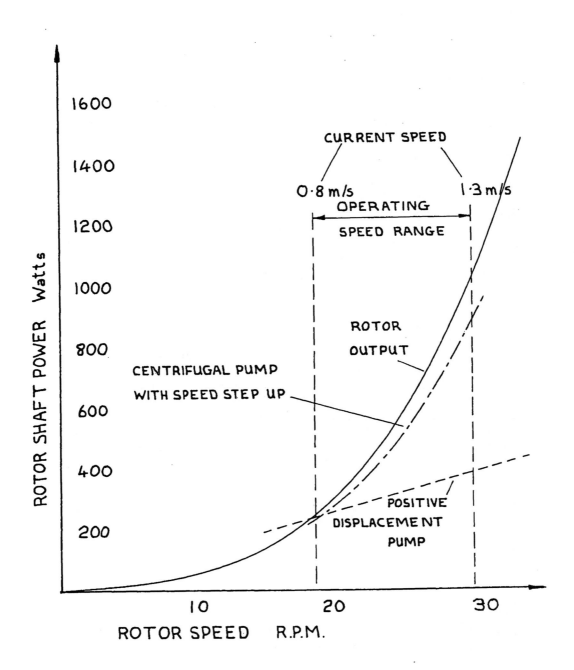

ROTOR AND PUMP POWERS vs ROTOR SPEED

FIGURE 2.13: Graph Showing Rotor and Pump Power Matching as a Function of River Speed.

2.7.2 Pump Selection for 'Mark 1' Type Machine

The SPP Unistream 40/13 centrifugal pump with a impeller diameter of 139 mm (see Figures 2.14 and 2.10) has been chosen as a suitable pump for total heads of up to about 12 metres. The correct transmission ratio varies between 40:1 and 100:1, depending on the rotor diameter, static head and pipeline dimensions, and is determined by the method described in Appendix 1. Two minor modifications are carried out to the pump. First the gland packing is removed from the stuffing box and replaced by a throttle bush, which allows a 0.25 mm radial clearance between it and the pump shaft. This removes the need for any maintenance of the stuffing box and is possible because the pump impeller is below the river surface. Second, the oil-lubricated pump shaft bearings must be replaced with grease-packed ones with seals. Because the pump shaft is not horizontal (see Figure 2.14) the bearing at the pulley end of the shaft would be starved of oil.

It is likely that any centrifugal pump with a 40 mm bore outlet flange and 140 mm (or thereabouts) diameter impeller will be suitable for the 'Mark 1' machine. Fitting a pump with a larger outlet will mean that the maximum head attainable will be reduced because it will have to be rotated slower to absorb the same power as the 40/13 pump. A pump with a smaller outlet will absorb less power at a given speed and therefore will be able to be run faster (ie at a higher transmission ratio) and hence generate a higher head. For total heads above 12 metres, the Unistream 32/13 pump (or similar) should be fitted and will give a maximum total head of about 21 metres. To generate this head a tansmission ratio of 114:1 would be required. If a 32 mm diameter pulley could be fitted to the pump shaft this would be no problem. Fitting a pump with a larger impeller would not necessarily result in a higher head being developed because if the, pump had the same outlet size it would have to run slower to absorb the same power. To be sure of the system generating more head the pump should also have a smaller outlet diameter.

The above paragraphs are only intended to give a rough idea of the type of pump to choose. If a manufacturer's pump performance curve (head vs discharge and power vs discharge) is available, it is possible to check its suitability for the turbine at a given site and then calculate the transmission ratio required and the pumped water output by the method given in Appendix 1.

FIGURE 2.14: 'Mark 1' Pump and Rotor Shaft in Operation.

FIGURE 2.15: Locally Fabricated Pump for 'Low Cost' Turbine.

2.7.3 Pump Selection for 'Low Cost' Machine

In spite of the apparently more complex concepts involved in
centrifugal pumps it is perfectly feasible to fabricate a
reliable open impeller pump from readily available mild steel
sections in a metalworkshop equipped with a lathe, arc
welder, pillar drill and hand tools. A pump efficiency of
over 35 per cent was achieved at the first attempt and in a
workshop with aluminium casting facitlities a more efficient
(and cheaper) pump could be made. The pump shaft rotates in
the same type of pillow block mounted ball bearings as used
for the turbine rotor shaft (see Figure 2.15).
    A discussion of the design of open impeller pumps is
unfortuantely beyond the scope of this book but the two
important factors to bear in mind are that the head generated
can be increased by increasing either the impeller diameter
or the rotational speed of the pump and the delivery is
increased by increasing the pump inlet diameter. The 'Low
Cost' design has an impeller diameter of 150 mm, an inlet
diameter of 35 mm and a rotational speed of 1,400 rpm. This
gives a maximum delivery of 1.5.1/s at 7 metre total head.
Under these conditions the pump will require about 300 Watts
to drive it, which is equivalent to the output of a sprint
racing cyclist. This, therefore, represents the maximum
power the turbine transmission (made from cycle components)
is designed to transmit. Increasing the size of the pump
significantly would increase the power required and result in
faster transmission wear (particularly on the cycle tyre –
see Section 2.6).
    Care is required in the manufacture and assembly of the
pump if a reasonable efficiency is to be obtained. The
diffuser (see Figure 2.12) is particularly important and
should ideally have a circular cross section with a taper of
less than 10 degrees along its length.

2.8 Delivery System

The delivery system is the means of getting the water from
the pump diffuser to the crops and may include steel and
polythene pipes and probably some earth channels. Although
the vertical distance the water has to be lifted is likely to
be only a few metres, the plant being irrigated is likely to
be between 25 and 100 metres away from the pump. The water
has to be transferred to the river bank in a pipe but there
are a variety of methods to get it from there to the plant,
each involving different capital costs and water losses.

FIGURE 2.16: Prototype 'Low Cost' Turbines at Prisons Gardens.

2.8.1. Pump to River Bank Pipe

Due to unsteadiness in the river current, the turbine will move in and out slightly in relation to the bank and so some flexibility must be built into this part of the delivery system. Originally, a solid steel pipleline on an oil drum float was used with flexible connections at each end, but this proved awkward to handle and was discarded in favour of the system shown in Figure 2.16 which uses rubber lined canvas fire hose suspended between the turbine and the bank. If a fire hose is used, care must be taken to avoid sudden bends which cause a restriction in the pipe and can seriously affect the quantity of water pumped. A suitable alternative is a thick walled polythene pipe which is usually more expensive but less liable to kink. It is advisable to try to keep the pipe out of the water at least in the main current to avoid drag on it which tends to pull the turbine in towards the bank and also to avoid weeds being caught on it.

Whatever method is used it is essential that the pipe is securely fastened at both ends and that on the river bank there is a length of steel pipe firmly anchored into the ground. If the pipe is not securely anchored at this point the whole delivery system may be pulled into the river when the machine is stopped.

Turbines have been tested at distances of up to 15 metres from the bank but with careful pipe arrangements it is probable that water could be piped up to 25 metres between turbine and river bank without significant problems.

2.8.2. River Bank to Plants

The choice of distribution system is an important consideration and it should be planned as far as possible to use methods already in use which are fully understood by the smallholders. Provided the gradients are favourable the cheapest method of distributing water to the various plots is by earth channels as shown in Figure 2.17. However, at the test site the losses through soakage were found to be of the order of one l/s per 100 metres of earth channel, so that, particularly with the 'Low Cost' machine, a large percentage of the machine's output can be lost resulting in a poorer financial return from the machine. Distribution through a flexible polythene pipe is much more efficient method, provided that its diameter is sufficient to keep pipe friction losses to an acceptable level. This point is discussed in more detail in Appendix I but a 2" bore pipe should be used for the 'Low Cost' machine and a pipe of a minimum bore of 2 1/2" used for the 'Mark 1' machine. Due to the difficulty of manhandling a long length of flexible pipe

FIGURE 2.17: Distributing Water Using Earth Channels and Flexible Pipe.

over plots full of vegetables, it is likely that some combination of pipe and earth channels will prove to be the best solution for the delivery system. Lining the channels with plastic or clay would reduce the soakage losses but would be more work to build. It is possible the reduce the wastage in the channels by planting vegetables or fruit trees on the earth ridges on either side.

2.9 Floats

The floats must provide enough buoyancy to hold the turbine in the correct position when running and to support the weight of two people working on the machine without allowing the transmission to get wet.

To keep costs to a minimum the 'Low Cost' machine is floated by empty 200 litre oil drums, and the absolute minimum number of drums required is four. With four drums care has to be taken that two people do not go near the transmission end of the pontoon at the same time as it will become submerged.

The life of the drums before rusting through is 18 months to two years, depending on how well they are painted. The drums cost between S£8 and S£10 in Juba market. Large hardwood logs, if cheaply available, might be used as an alternative for pontoon floats.

The 'Mark 1' machine is floated on ferrocement floats (see Figures 2.4 and 2.18) which, owing to their cost of about S£500 each, cannot be justified for the 'Low Cost' machine but have the following advantages over drums:

1. Given correct construction and curing their life will exceed that of every other part of the machine and it should be possible to re-use them when the mechanical parts are replaced.

2. The keels can be built into floats, so that it is not necessary to have rudders attached to the frame (see Figures 2.4 and 2.6).

3. Each float forms one side of the pontoon, so the only additional materials required are two cross timbers which are clamped to the floats.

4. These floats produce a very stable pontoon which can easily take the weight of four people as well as the machine.

5. The turbine's appearance is greatly improved by the use of purpose-made ferrocement floats.

Each float has two bulkheads built into it to form three separate watertight compartments in case of damage.

FIGURE 2.18: Plastering Ferrocement Float for 'Mark 1' Turbine.

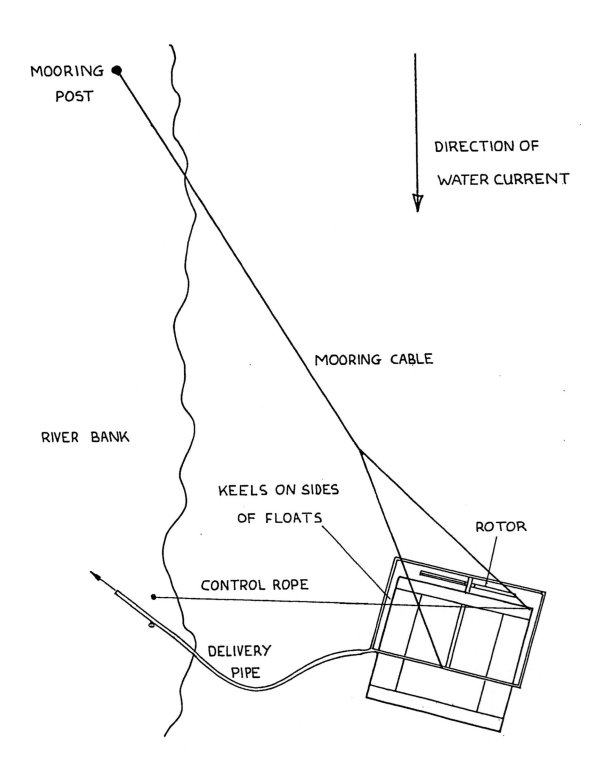

MOORING POST

DIRECTION OF
WATER CURRENT

MOORING CABLE

RIVER BANK

KEELS ON SIDES
OF FLOATS

ROTOR

CONTROL ROPE

DELIVERY
PIPE

FIGURE 2.19:   Sketch of Mooring System.

Ferrocement is widely used in the Third World as a material for water tank construction and somewhat less widely used for boatbuilding. Skills required are welding and plastering plus a good understanding of the important curing process. It is recommended that anyone without previous experience of ferrocement, practices plastering some vertical test pieces to get the stiffness of the mix right and to develop a good plastering technique before attempting to build a complete float.

2.10 Mooring System

The mooring system keeps the machine out in the river current when it is running and allows it to be easily brought into the bank for maintainance. Earlier work in northern Sudan and the experience of the Danish Guides and Scouts at Rejaf has shown riverbed anchoring to be impractical. The system shown in Figure 2.19 has proved to be a satisfactory solution.

The reaction of the water on the keels or rudders provides the force necessary to keep the turbine out in the current. To provide the force the keels must be held at an angle to the direction of flow, and this is done by spliting the mooring cable ahead of the machine and attaching one end of the yoke each side of the centre of drag. By altering the relative length of the cables it is possible to adjust the distance of the turbine from the bank. The keel area must be at least as great as the rotor swept area to avoid the pontoon having to be set at a large angle to the current direction which would decrease the effective rotor swept area.

If the cables are arranged as shown in Figure 2.19 the turbine can easily be pulled into the bank by means of the control rope. Pulling it will move the keels parallel to the current and the machine will drift gently in towards the bank. If the water near the bank is shallow it will be necessary to lift the rotor first. The machine is returned to its position in the river by simply pushing the upstream end of the pontoon out into the current.

2.11 Rotor Supporting Frame

The plane of rotation of the turbine rotor must be maintained in the correct position relative to the current flow. This is accomplished in the two designs under discussion by a rigid frame made from 50 mm bore galvanized steel pipe which supports each end of the rotor in a suitable bearing.

This arrangement provides a means of attaching the rotor to the pontoon and affords some measure of protection against accidental grounding for the blades. The materials, however, are expensive and one of the advantages of the trailing rotor machine discussed in Section 2.5.1. is that it does not require any frame.

The rotor frame is located on the pontoon by means of wooden bearings which allow it to pivot about the top frame tube so that the rotor may be raised for inspection.

2.12   Rotor Lifting Mechanism

Some means of lifting the turbine rotor out of the water is necessary for cleaning and maintenance and also to bring the turbine in to the river bank.

As long as the rotor can be rotated when it is out of the water it is not necessary for all three blades to be completely visible at one time and so the system used raises the rotor until the bottom bearing is clear of the water. Two small locally made winches are mounted on the pontoon and their cables attached to the corners of the rotor supporting frame immediately below them. The attachment points are well clear of the sweep of the rotor blade to avoid any chance of fouling if one cable were left slack and to avoid any weed which catches on them affecting the flow over the rotor. The winch is locked by simply straightening the handle and letting it rest on the pontoon front timber. Galvanized wire or nylon rope can be used for the winch cable, but in either case should be replaced yearly.

2.13   Storage Tank Design

One of the main differrences between a water current turbine and a diesel pump, as far as the user is concerned, is the rate of water delivery. Diesel pumps with comparatively low capital costs and high running costs are normally sized, so that between four and six hours running per day provides adequate water. The water current turbine, however, with its high capital cost and minimal running cost, should be run for as much of the day as possible to irrigate the maximum area and hence get the best return on the investment. In practice, irriation water cannot be applied during darkness or during the hottest part of the day and so the turbine would only be run for about eight hours. The economic comparison discussed in Section 3.4 and Appendix 2 is based on only eight hours running per day.

Running the turbine continously and storing the water in a tank during the night and middle of the day would have the following advantages:
1.   The vegetables would only be watered during the early morning and evening. This would save water lost through evaporation and be better for the plants.

2.   The rate of filling the basin with water could be controlled by a valve on the tank outflow and would not

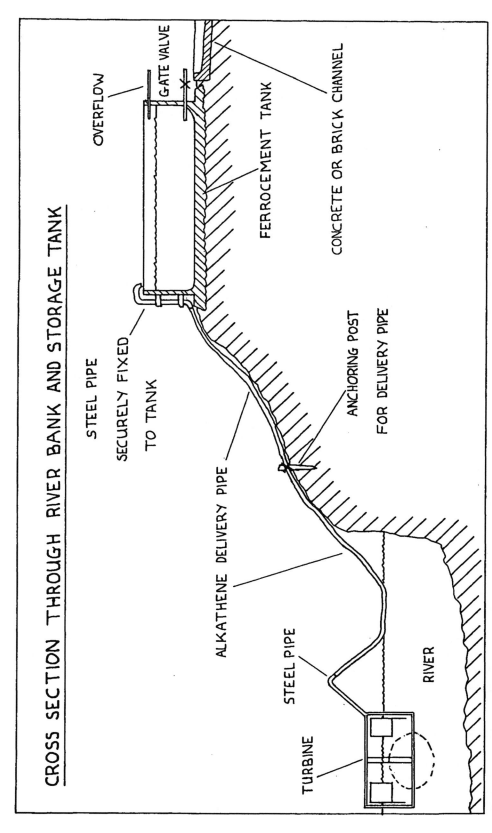

CROSS SECTION THROUGH RIVER BANK AND STORAGE TANK

OVERFLOW

GATE VALVE

STEEL PIPE

SECURELY FIXED
TO TANK

FERROCEMENT TANK

CONCRETE OR BRICK CHANNEL

ALKATHENE DELIVERY PIPE

ANCHORING POST
FOR DELIVERY PIPE

STEEL PIPE

RIVER

TURBINE

FIGURE 2.20;   Sketch of Storage Tank and Pipe Arrangement

47

be limited to the rate of delivery of the turbine. The basins would therefore be filled much faster and the labour required per unit area would be reduced.

3. Due to the rapid flow possible from the storage tank the percentage of water lost through soakage in an earth channel distribution system would be considerably less.

4. In the event of machine breakdown, whatever water was in the tank could be used to keep the most valuable crops alive until repairs were completed.

Building a tank which could store 10 hours of the machine's output should at least triple the area a given machine could irrigate. For the 'Low Cost' machine, the size of tank required would be 40 m³ which would cost about S£2,500 in southern Sudan to construct from ferrocement. With the necessary additional water piping the extra capital cost of water storage would be about S£3,100, bringing the total to approximately S£5,200, that is, 2.5 times the cost of the turbine without water storage. If the necessary capital were available, a tank would be a worthwhile addition to the system, particularly as it would require no maintenance other than occasional cleaning out and would last at least 30 years if built properly.

To avoid any chance of the tank draining back through the machine, the delivery pipe from the turbine should discharge over the lip of the tank as shown in Figure 2.20. The height of the tank is additional static head for the turbine to overcome and should therefore be kept to a minimum. At tank of 6 m diameter and 1.5 m height should be a reasonable compromise between materials cost and additional pumping head.

The construction of ferrocement tanks is described in detail in reference 10, p113.

2.14 Installation

2.14.1 Preparations

Before choosing a water current turbine as the appropriate pumping technology, the site under consideration must have been thoroughly surveyed to provide the information listed in Section 2.2. If after the survey and after consideration of the social and economic factors (see Chapter 3) it is decided to use a water current turbine, the following decisions must be made before manufacture starts.

(i) Whether to use the 'Mark 1' or the 'Low Cost' design (or a hydrostatic coil pump, see reference 9, p113). This decision will depend on the water output required, the materials and capital available and the minimum river speed and depth.

48

(ii)   The diameter of rotor to be used. This depends on the output required and the water current speed.

(iii) The transmission ratio required. This is calculated by the method described in Appendix I once the rotor diameter is settled and the static head and pipeline length and diameter known.

During manufacture of the machine, visits to the site should be made to establish the precise working position of the turbine and to install the mooring post. As already mentioned the water current speed can vary by up to 10 per cent a few metres up or down stream, so careful location is worthwhile. The site should be as free as possible from turbulence and eddy currents and there should be deep water as close to the bank as possible to allow for easy manoeuvering of the machine. Avoid rapidly eroding river banks and find a convenient place to bring the delivery pipe ashore.

The mooring post is sited upstream on a solid piece of river bank at a distance at least three times the distance required between the turbine rotor and the river bank. Thus, if it is necesary to site the turbine 15 metres out into the river to find a fast enough current speed, the mooring post should be about 50 metres upstream.

The mooring post may have to withstand forces of the order of one tonne so it must be substantial, preferably consisting of a 75 mm bore steel pipe, and well concreted in.

The post should be installed a few days before the turbine so that the concrete (which should be kept damp for at least a week) has reached a reasonable strength.

2.14.2 Assembly

The design of both machines is such that they can be transported in component form by road and assembled on site in the shallow water near the river bank. If it is more convenient the machines can be assembled up or down river and towed to the site. In current speeds of 1 m/s to 1.3 m/s a boat with engine power of at least 25 Horse Power will be required.

For the 'Mark 1' machine, six people will be required to manoeuvre the ferrocement floats down the river bank and a winch may be necessary where the bank is very steep. For the 'Low Cost' machine, three people are ample for assembly and commissioning.

Until members of the team have experience and confidence the assembly should be carried out in as near still water as possible, preferably about waist deep.

To start with, the mooring cable should be firmly fixed to the post and its free end attached to the first float before it is put into the water. No part should be put into the water until it is tied to the mooring rope as untethered objects will be lost if the current catches them.

Once assembly is completed, with the exception of the rotor blades and the delivery pipe, the functioning of the mooring system must be tested. If this is satisfactory the rotor blades are fitted and the mooring system again adjusted to keep the turbine in the correct position when it is running. The delivery pipe is then attached to the machine and firmly tethered where it reaches the river bank.

2.14.3 Commissioning

At this stage it is well worth carrying out as much performance measurement as is possible with the test equipment available. This is essential at the prototype evaluation stage to enable the cause of any subsequent decrease in performance to be traced.

Ideally, a Braystoke current meter (or similar), a stopwatch, a Bourdon pressure gauge and a tank of known volume are required, but even if the only test equipment available is a watch with a second hand and a 200 litre oil drum useful measurements can be made.

The following should be measured:
1. The water current speed just upstream of the turbine.

2. The rotational speed of the turbine rotor when delivering water through the complete delivery pipe system (time 300 revolutions).

3. The rate of water delivery (time the machine to fill a 200 litre drum).

4. The rotational speed of the turbine rotor when running with no load (remove the first belt or chain and time 30 revolutions).

5. The pressure at the pump outlet when the machine is delivering water.

Using the results from tests 1. and 2. and the information in Appendix I it is possible to check that the turbine is running at its most efficient speed relative to the current speed. If not, the transmission ratio should be altered and the tests repeated. Results from tests 1. and 3. enable the overall system efficiency to be calculated (see Section 2.1.4). A figure of 10 per cent for the 'Mark 1' or 7 per cent for the 'Low Cost' machine should be achieved given a reasonable pipeline efficiency. If it is lower than 75 per cent the use of a large diameter delivery pipe, or the addition of a second pipe in parallel to the first, should be seriously considered. The result of test 4. gives an indication of the efficiency of the turbine rotor (see Appendix I) and this will depend on the standard of workmanship of the blade construction and reasonable bearing alignment during assembly.

These tests should be repeated every three months or so during prototype testing, or whenever necessary to diagnose the cause of poor performance.

2.15 Operation

The ease with which water current turbines can be introduced will depend on a whole series of factors, some of which concern the day to day operation of the technology. In the field testing carried out in southern Sudan the following were found to be important:

1. Whether any water pumping device had previously been used at the site. If the turbine was simply a replacement for, say, a diesel pump the users were experienced in managing the established water distribution system and only had to adapt to the slower delivery rate of the turbine. If, however, the whole irrigation scheme was new and unfamiliar to the operators considerable time had to be put into their training, preferably by an agricultural extension worker.

2. The quantity of weed and debris in the river controls the amount of attention the turbine needs. Much effort has gone into designing the machines to catch the minimum amount of weed, but in some rivers the machines may require cleaning several times a day. Cleaning and restarting the machine is tedious, particularly in the heat of the day, and dangerous at night. Owner operators tend to be much better motivated to keep the machines running than employees but there will come a point where everyone will give up the struggle against the weed. Simple tools greatly ease the job of machine cleaning.

3. Attitudes to swimming in the river and any real or imaginary dangers involved will affect people's willingness to accept the technology. If the water near the bank is too shallow to bring the machine in to the bank with the rotor in the running position, the presence of snakes, crocodiles or a current speed of more than 1 m/s between the bank and the turbine will necessitate the purchase of a canoe or small boat, further adding to the capital cost of the system. The trailing rotor machine mentioned in Section 2.5.1 would not require a boat as it could be brought into shallow water without fear of damaging the rotor.

4. Proper training in manoeuvering the machine safely is essential. Owing to the large forces involved, getting the mooring cables tangled can cause serious accidents.

5. Stopping and starting of the machine is straightforward, but, again, training is necessary to establish safe working practices such as not leaving transmission guards off and always working downstream of a turning motor.

2.16 Maintenance

Maintenance consists chiefly of checking and adjustments with occasional replacement of winch cables on both machines, belts on the 'Mark I' machine and oil drums and bicycle tyre on the 'Low Cost' machine. A maintenance schedule for the 'Mark I' machine is shown in Appendix 1.5.

    If the 'Low Cost' machine is purchased by an individual smallholder, that person should be involved in the construction and installation of the machine and should thereafter be encouraged to be responsible for its routine cleaning and maintenance. If the small-holder is familiar with bicycle mechanics changing the cycle tyre should present no major difficulties and more skilled assistance will only be required annually to change the drums and winch cables and to check the machine over.

    The 'Mark I' machine is likely to be owned by a group of small holders or an institution of some sort, and so the responsibility for operating and maintaining the machine will have to be delegated to someone who is not the owner. If this person is inadequately trained or poorly motivated then there is very little chance of the turbine installation being a success, especially if there is a lot of weed in the river.

    Both machines are designed for easy assembly and dismantling, the various parts being fixed to the frame with U bolts rather than welding. Thus, if any part requires repair it can easily be unbolted from the machine and taken to the nearest workshop where the required tools are available.

2.17 Conclusions

2.17.1 Key Design Features

Over the four years of development and testing of water current turbines many alternative design features have been experimented with and many more discussed and researched. The following features are those which have made the machine a viable water pumping tool:

(i)   The mooring system which has enabled the machine to be moored to a single post on the bank and easily moved in and out of the current.

(ii)   The excellent match achieved between the rotor and pump which enables the machine to run efficiently in a varying current speed without requiring adjustment.

(iii)   The achievement of a reliable design (the 'Low Cost' machine) which can be manufactured from parts and materials locally available in most Third World countries and maintained by its owner.

(iv)   The capital costs of both machines are now low enough for them to an economically viable alternative to diesel pumps where fuel is expensive.

2.17.2 Main Technical Features which Affect Users

The testing work done with local farmers and the Juba Prisons Department gardens was invaluable in the development of the 'Low Cost' design. The difference between a machine which works when the designer and builder is there and one which will work reliably when left with a farmer was very clearly illustrated. The major criterion by which proposed design modifications came to be judged was whether they would make the machine 'user friendly' and safe to operate.

The following features were found to be important:

(i)   A self-priming pump is essential to make the machine easy to start up. This was achieved on both machines by arranging the pump with its impeller submerged.

(ii)   To reduce weed clearing time to a minimum the machine must be designed to catch the least amount of weed. Various deflectors and barriers were tried without success and so the number of pieces of metal or cable cutting the water surface (where the weeds float) was reduced drastically.

(iii)   The machine must provide a stable platform so that two people can work on it. This is particularly important when clearing weed from the machine.

(iv)   Ideally no tools should be needed in normal operation and checking of the machine, and a minimum number used to assemble and dismantle it.

(v)   The rate of delivery of the 'Low Cost' machine is very much less than that of a small diesel pump and is considered unacceptably low by some users. It was felt that a storage tank would have made each machine much more satisfactory because the rate of irrigation could be controlled by a gate valve to allow the farmer to flood the beds as quickly as required.

(vi)   At sites where, because of inadequate depth of water near the river bank, the machine's rotor has to be raised before it can brought in to the bank, it is necessary to either swim to the machine or own a boat or canoe. In either case it is quite a business to get onto the machine and enthusiasm soon wears thin if frequent visits are required to clear weed.

2.17.3 Further Development Work

As already stated, the work done so far is only a beginnning and further development and field testing should be directed towards the following aims:

(i)   reducing the capital cost of the machines. It is expected that the trailing rotor machine, (see Figure 2.9) once sufficiently tested, would be considerably cheaper than the 'Mark I' or 'Low Cost' designs,

(ii)   improving the performance of the 'Low Cost' machine by further work on the transmission and pump design,

(iii) further improving the ease of operation of both machines.

CHAPTER THREE

Socio-Economic Analysis

3.1 Introduction

There are three main aspects to the socio-economic appraisal of the water current turbines for irrigation. These are:

(i) economic analysis - to establish the maximum costs above which WCTs are unlikely to be economically attractive to farmers;

(ii) consideration of social factors; and,

(iii) if they do appear to be economically viable and socially acceptable, the systematic comparison of WCTs with alternative pumping technologies - to determine which system is likely to be the most socially acceptable and constitute the best value for money.

The sequence in which these question should be addressed is shown in Figure 3.1. Steps (i) and (ii) can proceed simultaneously. Step (iii), which is only relevant if WCTs pass the 'tests' set by steps (i) and (ii), can be conducted at varying levels of detail - here, we present the outline of a method for systematic comparison, some evidence of the circumstances in which alternative systems may be economically competitive with WCTs, and references for further study.

The order of this chapter reflects these priorities. The use of a simple economic decision rule, the payback period, to effect step (i) is explained in Section 3.2. Details of an example of this approach, based on the economics of vegetable gardens using water current turbines in Southern Sudan, are given in Appendix 2. Next, the wider social considerations of the alternative pumping technologies are discussed in Section 3.3. Finally, step (iii) is considered in Section 3.4. This last section summarizes the evidence on the relative cost-effectiveness of alternative pumping systems. Some further details are given in Appendix 3.

3.2 Assessment of the Cost-effectiveness of Water Pumping For Irrigation

The question most fundamental to the success of a new water pumping technology is: 'will it make money for the farmers?'

To answer this question we advocate the use of a simple economic decision rule - know as the payback period. In the case of irrigation, the payback period is the length of time

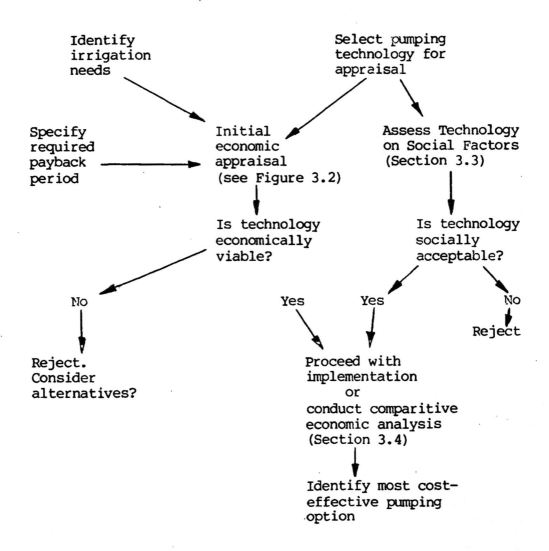

FIGURE 3.1   Main steps in socio-economic appraisal

required to pay back capital costs from annual profits of crop sales. The basic formula for estimating the payback period is as follows:

$$\text{Payback Period (PP)} = \frac{\text{capital cost}}{\text{annual crop revenue} - \text{annual recurrent cost}}$$

The steps in the calculation are:

(i)   talk to farmers to determine how soon they expect to earn a profit from investments;

(ii)   as a result of these discussions, specify the maximum acceptable payback period;

(iii)   collect information on the capital and recurrent costs of WCTs, and of expected annual crop revenues, to estimate the actual payback period;

(iv)   compare the estimated PP with the maximum period acceptable.

In implementing step (i), it is essential that extensive discussions with local farmers occur. Payback requirements will be conditioned by prosperity (particularly savings, if any, and capital possessions) and past experiences. For example, in areas characterized by severe drought every six to seven years a payback within two to three years is likely to be sought; in more moderate or more predictable climates, longer payback periods may be acceptable. A further important factor may be the existence (and terms) of credit provision - some of the consequences of this are discussed in Section 3.3.4.

The method of estimating the payback period for a WCT is shown in Figure 3.2. A specific example, of the economics of irrigating vegetable gardens by a WCT in Southern Sudan, is summarized in Table 3.1 and presented in more detail in Appendix 2.

In practice, as shown in Table 3.1, the payback formula can be used in two different ways. First, if the required payback period is specified (and annual crop revenues and recurrent costs estimated) the maximum acceptable capital cost can be estimated. Alternatively, all cost and revenue estimates can be input to the formula and the payback period calculated and compared to the required value. We recommend the first of these two approaches - because it forces fieldworkers, at the outset of the study, to determine what realistic acceptable payback periods are.

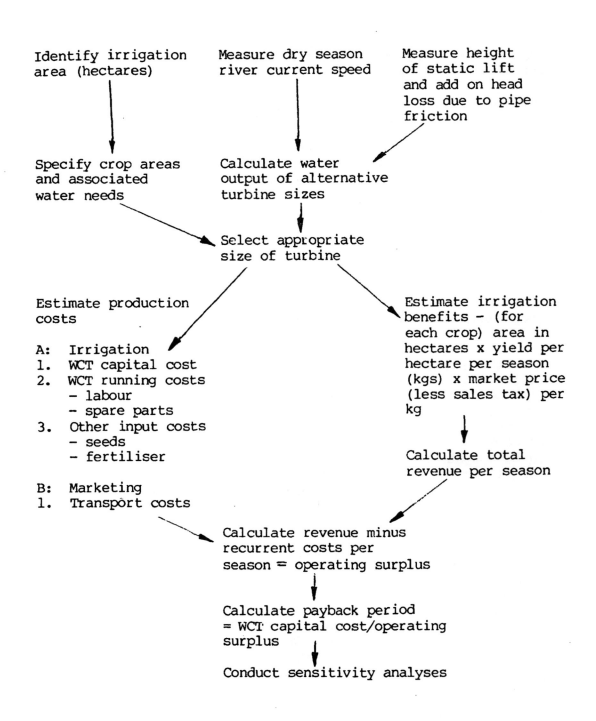

Identify irrigation area (hectares)

Measure dry season river current speed

Measure height of static lift and add on head loss due to pipe friction

Specify crop areas and associated water needs

Calculate water output of alternative turbine sizes

Select appropriate size of turbine

Estimate production costs

A: Irrigation
1. WCT capital cost
2. WCT running costs
    - labour
    - spare parts
3. Other input costs
    - seeds
    - fertiliser

B: Marketing
1. Transport costs

Estimate irrigation benefits - (for each crop) area in hectares x yield per hectare per season (kgs) x market price (less sales tax) per kg

Calculate total revenue per season

Calculate revenue minus recurrent costs per season = operating surplus

Calculate payback period = WCT capital cost/operating surplus

Conduct sensitivity analyses

FIGURE 3.2: Steps in estimating the payback period of a pumping technology for irrigation

	Base case	Sensitivity Analysis Test 1(2)	Test 2(3)
Required payback period (dry seasons)	3	3	3
Crop revenue per season (S£) (7)	2,000	2,000	1,000
Recurrent cost per season (S£)			
: Irrigation – fuel	0	0	0
: " – labour (1)	180	180	180
: Cultivation labour cost	360	360	360
: Other input cost (8)	100	100	100
Sub Total Recurrent Cost	640	640	640
Operating surplus = crop revenue Less recurrent cost	1,360	1,360	360
Maximum acceptable capital cost = required payback period times operating surplus	4,080	4,080	1,080
Actual capital cost ('Low Cost' Machine)	2,070	4,140	2,070
Actual payback period (dry season)	1.5	3.1	5.8
Policy conclusion (on these assumptions is technology economically viable?)	Yes (4)	No (5)	No (6)

Notes

(1)   Estimated irrigation labour cost
(2)   Sensitivity test 1 : doubling of actual capital cost
(3)   Sensitivity test 2 : halving of crop revenue
(4)   Payback in 1.5 dry seasons
(5)   But very marginal; payback will take 3.1 dry seasons
(6)   Payback will take 5.8 dry seasons
(7)   For estimation see Appendix 2, Table A2.1
(8)   Seeds, fertilizer, etc.

Source: Sudan data (Reference 2, p104) see Appendix 2

TABLE 3.1: Illustration of Payback Method Estimating Maximum Acceptable Capital Cost

3.2.1 Sensitivity Analysis

The calculation in Table 3.1 shows that for the 'Base Case' assumptions, the 'Low Cost' water current turbine pays for itself in one-and-a-half-dry seasons and thus can be considered to be economically viable. However, before a firm decision can be reached on economic viability, it is essential to conduct an analysis of the sensitivity of the estimated length of the required payback period (or the maximum acceptable capital cost) to variations in key input variables. These key variables will include:

(i)   achievable water output

(ii)   crop yields

(iii) crop prices (before and after the introduction of irrigation)

Examples of such sensitivity analyses - which are not difficult to conduct - are shown for the Juba case study in Table 3.1. The results of two tests are shown. Test 1, a doubling of the capital cost of water current turbines, shows that WCTs cease to be economically viable - but only marginally so; compared to the 'base case' assumptions the payback period increases from one-and-a-half to just over three dry seasons (actual capital cost is S£4,140 compared to S£4,040 required to achieve a payback in three dry seasons). Test 2, a halving of crop revenues, shows WCTs to be clearly not viable economically - actual capital cost is nearly double the maximum acceptable level and the payback period is almost six dry seasons.

3.2.2 Conclusion

This concludes our presentation of the method of applying the concept of a payback period to assess the economic viability of using water current turbines for irrigation. More complex methods of economic analysis (involving the discounting of future costs and benefits) can be used. We advocate the payback criterion in this context because:

(i)   it is straightforward to use and can be easily understood by fieldworkers and also by local people;

(ii)   it disregards costs and benefits beyond the required payback time - and we believe that, particularly in poor rural communities where many factors make future costs and benefits from irrigation highly uncertain, it is appropriate to assess the economics in this way, and;

(iii) a technology which yields an acceptable payback period is also likely to be economically

60

acceptable on more complicated criteria (for example, for a pumping system with an expected life of five years, a payback period of t h r e e years is equivalent to an internal rate of return (IRR) or 20 per cent, and a payback of four years to an IRR of 8 per cent.

Ideally, if WCTs satisfy the payback criterion, alternative pumping technologies should be analysed in a similar way to determine which system constitutes the best value for money. Some initial guidelines, upon the circumstances in which each of these alternatives may merit consideration, are given in Section 3.4. However, the need to consider alternatives will depend on the likely social acceptability of WCTs - and so it is to a consideration of the social issues that we now turn.

3.3 Social Factors

3.3.1 Introduction

The primary concerns to a potential purchaser of a water current turbine for irrigation will be whether it will be a profitable investment (as discussed in Section 3.2) and furthermore (as discussed in 3.4 below), whether, of the range of alternatives, it constitutes the best available value for money.

In addition, however, there are wider social criteria which must be satisfied if the investment is to be successful. Significantly, the alternative pumping technologies score differently on these various social considerations - and so factors must be carefully considered in technology appraisal. At the outset of this discussion it is important to recognize that the purpose for which water is to be pumped has an important bearing on the relative importance of the various social factors. In this respect, the key characteristics of water pumping for irrigation are:

(i) ownership of pumping systems will often be private - but may be communal, depending on the optimal size of available pumps relative to the typical size of land-holdings; and

(ii) the costs and benefits of introducing pumps will be mainly financial[1].

---

[1] These characteristics contrast, for example, with pumping for village water supplies - where ownership of them is communal and schemes primarily generate non-financial benefits.

61

In the remainder of this section, we draw on these key considerations, together with other evidence, to identify the most important social criteria to be considered, and compare WCTs with the alternative pumping technologies according to these criteria. This process illustrates the proposed method of social appraisal. A full checklist of factors is given in Chapter 4. The key social criteria are:

(i)   size of pumping system relative to size of landholdings;

(ii)   constraints imposed by type of pumping system on the rate of water output;

(iii) capital costs (and associated need for credit provision);

(iv)   effects of constraints on pumping siting on land values, and;

(v)   potential for local manufacture.

In addition, a discussion of the purely technical features of water current turbines which will influence user acceptability (for example, through the skill levels required for operation and maintenance) is given in Section 2.17.2.

3.3.2 Size of Pumping System Relative to Size of Landholdings

The field experience of WCTs in southern Sudan illustrates the problems which may arise when the size of the area which a single pump can irrigate exceeds the typical size of landholding – so that several small-holders are involved. An example of <u>compatibility</u> between pump size and area under single ownership is the vegetable garden owned by one of the Juba Boatyard workers, Marco Oping, and operated by his family. This garden, is irrigated using a small water current turbine (output 1 litre/second). Communal operation involving several people but a single family – ensures that there are no disputes about who should get how much water at what time of day. In contrast, exactly these types of dispute did occur in the larger (3 hectare) small-holder scheme where one of the larger turbines was installed. This latter enterprise involved 10 small-holders with little experience of communal irrigation management.

More generally, the evidence collected in the Juba area shows the size of small-holdings to be well suited, particularly to the smaller 'Low Cost' WCT. A survey of growers on the West Bank (or Juba side) of the Nile (Reference 2, p113) shows 70 per cent of gardens to be less than 2 feddan (0.8 hectare) in area and thus compatible with the output of the smaller machine (See Appendix 1.3).

The size characteristics of WCTs are compared to those of the alternative pumping technologies in Table 3.2. The main points to note are:

1. The smallest size of diesel pumps widely available in poor countries (typically about 5 hp) is capable (assuming a lift of less than 5 metres) of irrigating an area of up to 4 hectares. In many countries this is likely to exceed the size of landholdings[1] of the majority of farmers. Thus, either diesel pumps are used at less than maximum capacity or arrangements for communal use must be evolved.

2. In contrast, handpumps are available to irrigate areas of less than half a hectare - typical of the smallest landholdings - so that, if they are introduced, indivisibility is not an important problem.

3. The two designs of water current turbine, developed at Juba, are capable (through a lift of 5 metres) of irrigating areas which are intermediate in size between hand and diesel pumps. The 'Low Cost' machine is appropriately sized for the majority of farmers; in the case of the 'Mark I' machine, there may well need to be consideration of whether cost-effectiveness will necessitate communal ownership, and what this will imply.

4. The size characteristics of WCTs, solar, wind and animal-powered systems - in terms of their divisibility - are similar.

---

[1] The appropriate measure is the average size of 'effective' landholdings in adjacent fields. The measurement of 'effective landholdings reflects ownership and operation, eg a farmer operating 10 acres under a 50:50 tenancy arrangement has an 'effective' landholding, contributing to this personal incomme, of 5 acres - and the size of pumping system appropriate for his needs should take account of this factor.

Part (i): optimal number of pumps to irrigate a 2-hectare
plot

Pump Type	Number
Diesel	1
Solar	2
Wind	2
Animal	2
Handpump	10

Part (ii): areas which can be irrigated by proven water
current turbines under typical conditions

Mark I Machine	3	ha
Low Cost Machine	0.5-0.75	ha

TABLE 3.2: Comparative Sizes of Alternative Pumping
Technologies

3.3.3 Restrictions on Water Output

The alternative systems for water pumping differ
significantly in the constraints which they impose upon users
in terms of the daily pattern of the availability of water.
These constraints may have an important impact on social
acceptability, depending particularly upon the role of
irrigation and the water requirements of the crops grown.
The three main types of irrigation practice are:

    (i)   irrigation of crops which could not be grown
    otherwise;

    (ii)   supplementary irrigation to increase yields;

    (iii) 'life saving' irrigation - to bridge a gap in
    rainfall during the wet season.

The most important distinction in crop types is between
crops which require regular application of relatively small
volumes of water and those which flourish in response to the
less frequent supply of larger quantities.

The contrast between the alternative pumping systems on
this criterion essentially derives from the source of the
energy harnessed to pump the water. The alternative systems
can be divided into the following categories.

    (i)   systems based on renewable energy sources - such
    as water current turbines, wind and solar pumps;

(ii)   systems based on human or animal power;

(iii) systems based on fossil fuels.

According to the local environment, all energy systems involve some degree of unpredictability. The energy available to WCTs and wind pumps is a function of the cube of the water current and wind speed respectively. Thus, a small increase in current or wind speed produces a large increase in water output (see Section 2.1.2 and Figure 3.3). At any given site, wind speed will vary between zero and gale force with a mean somewhere in between. Clearly, during project appraisal adequate data must be collected to determine a realistic estimate of average wind speed and the variation expected during the pumping season. Wind pumps nearly always require a water storage tank whose capacity is dependent on the length of the maximum probable calm period. Water currents, are however, much more predictable and, unless the river dries up completely, there is no 'calm period' to worry about. In practice, on the White Nile, variations in water current speed were found to be low and fell into two categories: one being a 'short term' variation (of period typically between one and three minutes and amplitue typically ± 5 per cent of the mean, depending on the site) due to unsteadiness in the current, and the other being a gradual decrease in mean current speed over the dry season (in the order of 15 per cent). It was therefore possible, with very few measurements, to determine the minimum likley current speed at any site, and if this figure is used to size the turbine there will always be at least the required amount of water available every day. Any water storage is then only short term (eg overnight) and purely to make irrigation easier and more efficient.

Water output from solar powered systems follows a regular daily pattern - the maximum rate being achieved when the sun is at its highest point during the middle of the day. This may have important effects upon user acceptability, particularly as the middle of the day is the hottest time and that least suitable for arduous agricultural work! (For a further discussion of this issue with respect to solar and wind powered systems, see references 3 and 11, on p113).

Pumping systems based on human and animal power, in principle, can provide water when required. In practice, supply restrictions may arise due to competing demands for agricultural labour. This very important issue is not a constraint arising from the technology per se but rather relates to the value which should be placed on agricultural labour. Rural communities in regions which experience highly seasonal climates, especially those engaged in fallow farming, arable irrigation farming or a combination of agriculture and livestock production, are particularly likely to attach high values to savings in labour. (For further discussion of these issues see references 4, 14 and 15, on p113 and p114).

WATER OUTPUT vs CURRENT SPEED

FOR 'MARK I' TURBINE

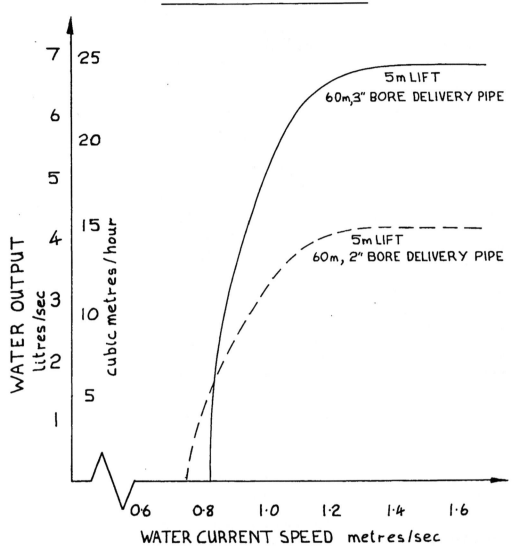

FIGURE 3.3: Graph Showing Discharge as a Function of River Speed for 'Mark 1' Machine.

Fossil-fuel based systems, in principle, render water available on demand to the farmer. In practice, particularly in remote areas, fuel shortages may be the norm or prices so high that supplies must be reserved for premium uses.

To conclude, the reliability of the intended power source must be very carefully assessed when comparing alternative pumping technologies. The provision of storage tanks can reduce this problem, but these will involve a significant additional cost[1] and (in communal schemes) will generate operational problems - in the distribution of water between users. In terms of the three types of irrigation practice identified above, renewable energy based systems are most suited to providing supplementary irrigation and least suited to 'life saving'. In environments such as the southern Sudan, where it is not very windy[2] and handpumping or watering from solar pumps in the midday sun may not be popular, taking water output into consideration, WCTs are particularly suitable.

3.3.4 Social Implications of the Differing Cost Structures of the Alternative Systems

Indicative estimates of irrigation costs are shown in Table 3.3. Part (i) of this table is based on 'international' cost data drawn from information from Kenya, Thailand and Bangladesh (see reference 3, p113). Part (ii) relates to southern Sudan. These two sets of data are not directly comparable - but do give an indication of capital costs, and of the contrast in the relative importance of capital and recurrent costs between the alternative systems.

The estimates in Part (ii) of the table are the costs of irrigating a larger area through a higher lift than the area and height of lift irrigated by the expenditures shown in Part (i). Thus, in terms of capital costs, water current turbines are less than one third the capital cost of solar powered systems and less than half the capital cost of wind power. Compared to the capital costs of other systems, WCTs are 70 per cent more expensive than diesel pumps (on the southern Sudan evidence) and similar in capital costs to animal pumps and handpumps.

---

1 Reference 3, p113 states these costs (at 1982 prices as US$ 58 per cubic metre; storing half the daily water requirements for a 2 hectare plot will require a tank of at least 40 cubic metres.

2 Defined very roughly as where there is not a fairly constant wind of at least 2.5 m/sec.

Part (i): Costs of irrigating 2 hectares through a lift of 2 metres ('International' cost data; analysis period 30 years)

System	Capital Cost	Total Discounted Lifecycle Cost (LCC)	Capital Cost (as % LCC)	Recurrent Cost (as % LCC)
Solar	17,060	19,602	87	13
Wind	11,070	13,012	85	15
Diesel				
: Low (40c/litre)	2,470	9,644	26	74
: High (.70c/litre)	2,470	21,068	12	88
Animal	3,630	12,391	29	71
Handpump	3,755	31,737	12	88

Part (ii): Costs or irrigating about 3 hectares through a lift of 5 metres (southern Sudan cost data; analysis period 10 years)

System	Capital Cost	Total Discounted Lifecycle Cost	Capital Cost (as % LCC)	Recurrent Cost (as % LCC)
WCT: 1 m/sec	4,950	11,610	43	57
1.2 m/sec	4,950	7,854	63	37
Diesel: (55 c/litre)	2,915	8,085	36	64
($3.2/litre)	3,443	21,593	16	84

1. The second column relates to total lifecycle costs, excluding replacement costs, discounted at a rate of 10 per cent

2. Sources: Part (i) reference 3, p113 (Table 8.7) (for underlying assumptions see Appendix 3)

   Part (ii) reference 2, p113.

TABLE 3.3:   Cost of Irrigation Using Alternative Pumping Systems (1982 US$)

In terms of recurrent costs, the most striking contrast (which has important social implications) is between the renewable energy-based systems (water current turbines, wind and solar power) and diesel pumps[1].

Recurrent costs are less than 15 per cent of the total discounted lifecycle costs of solar and wind powered systems. In the case of WCTs, the main recurrent cost is the labour time of pump attendants. This cost is a function of the rate of water output and hence of water current speed; at a speed of 1.2 m/sec total recurrent costs are 37 per cent of total discounted lifecycle costs. In contrast, depending on the price of fuel, recurrent costs of diesel systems represent between 64 and 84 per cent of total discounted lifecycle costs.

The social implications of these contrasting cost structures arise from the different problems associated with funding capital and recurrent costs. The main points to note are:

1. As a broad generalisation, aid agencies are much more willing to fund capital than recurrent costs.

2. In many poor countries the available funds to meet recurrent expenses are grossly inadequate: this problem is particularly acute in remote regions and when foreign exchange is required.

3. Typically, rural credit facilities are poorly developed, may involve extortionate rates of interest or may be viewed apprehensively by the local people.

An acute example of point 2. is provided by the Sahel region (references 4 and 12). Public sector deficits in seven Sahelian countries, projected for the period 1982-1984 average 20 per cent of forecast revenues. In practice, these gaps are reduced to zero by advancing aid disbursements, rescheduling debt repayments and, importantly, curtailing 'essential' recurrent expenditure. In these circumstances, even if the sale of crops generates funds which can be used for fuel purchase, shortage of foreign exchange and distributional problems may mean that supplies are not available. Even if fuel can be purchased, an additional social impact is the need for organization to collect the required funds - and to ensure that those who benefit pay.

---

[1] The high recurrent costs of animal and handpump systems are due to maintenance requirements and the value placed on attendance labour. The social implications of this latter factor are discussed above in Section 3.3.3.

An example of the problems associated with rural credit provision (point 3. above) is provided by handpump programmes in Bangladesh (reference 13, pll3). Complex certification requirements, the use of land as collateral and poor dissemination of rule changes to bank managers caused severe delays in programme implementation. As noted elsewhere (see Section 3.3.2) handpumps are the smallest (ie the most divisible) of the alternative pumping technologies; in a given social and economic environment, the problems of providing the larger capital sums required for the purchase of solar and wind pumps and, (to a lesser extent) water current turbines, may be more serious.

We conclude from this discussion that, in terms of the social requirements imposed by their cost structure, water current trubines compare well with the alternative pumping systems. The main advantages are:

(i)   WCTs have low recurrent costs - particularly on items requiring foreign exchange expenditure - a major advantage over diesel;

(ii) the capital costs of WCTs are low compared to solar and wind-powered systems.

The only serious disadvantage, on this criterion, is that the introduction of WCTs (particularly the larger 'Mark 1' version) may entail greater problems associated with the organization of credit provision than more divisible, smaller, systems - such as handpumps.

3.3.5 The Effect of Constraints on Pump Siting on Land Values

Water current turbines require that the pump be located in a river. The siting of the alternative systems is also constrained - but not in so clear-cut a fashion (for example, other things being equal, wind pumps should be located in the windiest locations); generally the siting of the alternative systems may involve trade-offs between a variety of factors - such as the energy available (and required) to raise water compared to that needed to pump to the most fertile areas.

A consequence of this characteristic of WCTs is that their adoption will place a premium on the value of agricultural land immediately adjacent to the river. This may be an important point against WCTs if there is a shortage of suitable land available. Alternative technologies may impose other land constraints - notably if the height of the water table falls markedly as distance from the river increases, making pumping uneconomic - particularly when using handpumps (see Figure 3.4 in Section 3.4 below).

70

3.3.6 Potential for Local Manufacture

In addition to cost considerations, there are two important advantages of a high degree of involvement by local people in the manufacture and installation of a new technology. These are:

(i) development of commitment and local skills - so that local staff are motivated and better qualified to under-take maintenance; and

(ii) generation of local employment and income.

3.3.7 Evidence of the Potential for Manufacture of Water Current Turbines in Southern Sudan

The two versions of the WCT differ in their potential for local manufacture. The 'Low Cost' machine incorporates a locally-made pump and transmission mechanism - whereas, for the larger 'Mark 1' machine, these items have to be imported.

An approximate breakdown of the total costs of the 'Low Cost' machine, into local and foreign exchange components, is shown in Table 3.4. Local costs account for some 62 per cent of the total - so that, in the manufacture of each machine, assuming a total cost of US$2,000, nearly US$1,240 is paid locally. If, as an example, the recipients spend 40 per cent of this income locally and this percentage of income is spend locally on each subsequent circulation of this income, the total income generated (the 'multiplier' effect) is approximately US$2,066. It should be noted that, in countries with appreciable manufacturing industries, the proportion of local costs would be significantly higher.

In contrast, the local expenditure component of the larger 'Mark 1' machine is lower (of the order of 35 per cent).

3.3.8 Comparison with Alternative Technologies

Clearly, the potential for local manufacture will vary between locations according to the raw materials and skills available. In addition, each of the alternative pumping systems has a variety of design options - so that, without a detailed study, only very broad general statements are possible. The main general points which can be made are:

(i) that solar powered systems clearly offer less potential than WCTs for local manufacture; solar modules (which account for 50-60 per cent of total solar system costs) are a new complex technology requiring expensive equipment to manufacture. Similarly, there are unlikely to be the facilities

71

to produce the other principal components of solar powered systems (electric motors, pumps and array support structures) in many of the poorest developing countries;

(ii)   for similar reasons, WCTs are more promising than diesel pumps for local production; and

(iii)   the manufacture of WCTs requires similar facilities, materials and skills as those needed to make steel wind pumps (such as the IT wind pump now in production in Kenya and Pakistan). In the case of handpumps, progress has been made and extensive efforts continue in many countries to develop designs capable of village level operation and maintenance - and, in some cases, local assembly or manufacture. Again, similar inputs to those required to manufacture WCTs are needed.

We conclude from this brief discussion that WCTs are likely to compare favourably to solar and diesel powered systems in terms of their suitability for local manufacture and are of similar suitability to wind or handpumps. A list of skills, facilities and materials required for manufacture of WCTs is given in Section 4.3.2.

3.4   The Relative Cost-Effectiveness of Water Current Turbines For Irrigation Compared with Alternative Pumping Methods

3.4.1 Introduction

In this section we assume that a payback calculation (outlined in Section 3.2) has shown water current turbines to be economically viable and that social surveys suggest that they will be acceptable on these criteria. The next question is : are WCTs the cheapest pumping method?

The answer will depend on the values of certain key site specific physical and economic parameters. What we can do here is to present some evidence on the costs of using WCTs for irrigation compared to the costs of using alternative pumping technologies. The purpose of presenting this information is to provide some initial evidence on the circumstances in which each of the alternative systems may be competitive with WCTs - and so should be subjected to the payback analysis described in Section 3.2.

Water current turbines are a newly developed technology for which little evidence of economic performance is available. The only source of information is the Study (Reference 2) based on information from Southern Sudan which compares WCTs to diesel pumps. In addition, we present evidence from a recent detailed study (Reference 3) which calculated and compared the unit costs of pumping water for irrigation using diesel pumps and various alternative

% Total Cost

Cost Item	Local Cost	Foreign Exhange Cost
Materials	27	22
Labour : skilled	14	-
: supervision	5	-
Overheads	16	16
TOTAL	62	38

Notes

(1).   Labour costs 72 man days, based on prototype construction time.

(2)   Overheads are approximately half local currency (rent of workshop, watchmans wages, workers welfare etc) and half foreign exchange (fuel and spares for generator, expatriate salary etc).

(3)   Estimated total cost US$2,000 (at 1982 prices).

Source : Reference 2, p113.

TABLE 3.4:   Local and Foreign Exchange Costs of 'Low Cost' Water Current Turbine (in Southern Sudan).

technologies. By combining the results from these two studies, we can make some general comments on the relative cost-effectiveness of WCTs. These remarks should be treated with caution - local information related to the key physical and economic parameters must be collected[1].

These sources of information compare the alternative pumping systems on the basis of unit water costs. The main steps in the estimation of these costs are described in Appendix 3.

3.4.2 Evidence on Comparative Irrigation Costs Using WCTs, Diesel Pumps and Alternative Pumping Systems

The evidence in Appendix 3 shows that:

    (i) under specified 'baseline conditions, diesel pumps are cheaper than alternative pumping methods; and

    (ii) in the southern Sudan, at current speeds in excess of 1.2 m/sec, water current turbines are cost-competetive with diesel.

Thus, at high current speeds, WCTs may well be the most cost-effective pumping technology. The question then arises : under what circumstances (assuming these current speeds) is reach of the alternative systems particularly likely to offer better value?

Figure 3.4, based on the assumptions in Appendix 3, illustrates point (i) above and provides some basic evidence on this issue. At high current speeds (of at least 1.2 m/sec), for the reasons stated in Appendix 3, WCT costs are similar to those shown by the line 'Diesel (low)'. Figures 3.5, 3.6 and 3.7 show the sensitivity of the unit costs of water pumping, using wind, solar and handpumps to variations (from the 'baseline' assumptions) in wind speed, solar irradiation and wage rates respectively. From this limited evidence, the main points on the cost-competitiveness of the alternative systems to note are:

---

[1]   A detailed handbook, outlining the steps which fieldworkers should follow to conduct a technical and economic appraisal of solar powered systems compared to wind, diesel, animal and handpumps, is available, see reference 11, p113.

COST vs HEAD

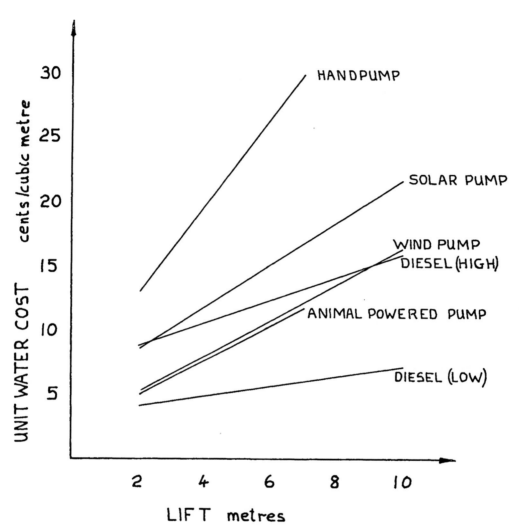

SOURCE: REFERENCE 3

FIGURE 3.4:   Comparative Costs of Water Pumping Using
Alternative Systems.

1. Diesel systems are a clear candidate for economic comparison with WCTs. However, it should be emphasized that, particularly in the very poorest countries (such as the Sahel region, (reference 12, p113) funds for recurrent expenditure - such as the purchase of diesel fuel - are in very short supply. In consequence, there is a strong case for attaching a weighting (g r e a t e r than one) to recurrent costs in project appraisal. The evidence from Southern Sudan illustrates this point well; there is a good economic case for attaching a higher cost to diesel than the official price - such as the unsubsidized price in Appendix 3. A weight of this magnitude makes WCTs considerably more attractive than diesel.

2. Handpumps on the 'baseline' assumptions (shown in Figure 3.4) have much higher unit water costs. The cost-competitiveness of handpumps decreases markedly as the height of lift increases, and, as shown in Figure 3.7, handpumps are a much more economically viable alternative to WCTs if no cost is attached to pumping labour. In general, and particularly during periods of peak agricultural activity, it is likely that labour will have a real economic value. Handpumping sufficient quantities of water for irrigation is a very time and energy-consuming activity. For example, handpumping to 60 cubic metres of water (a typical peak daily requirement to irrigate one hectare) through a lift of 2 metres requires three people each working for seven hours - and twice this number of people are required if the height of lift is 4 metres.

3. Animal Powered Systems, on the 'baseline' assumptions, merit consideration as cost-competitive options to WCTs. As with handpumps, the value of labour is a key factor.

4. Wind Powered Systems are also a potentially cost-competitive alternative. However, as shown in Figure 3.5, the unit water costs of these systems are highly sensitive to variations in wind speeds. If the intended location is thought to be characterized by relatively constant winds in excess of 3.5 m/sec, wind power may be economically more attractive than WCTs. A guide to the required appraisal procedure is shown in reference 11 on p113.

5. **Solar Powered Systems**, at present day solar module
    costs, do not appear to be a strong alternative, on
    economic considerations, to WCTs. However, solar module
    costs are projected to fall - and are on course to reach
    the 'target' costs `shown in Figure 3.6 by 1987.   If
    these 'target' costs are achieved, solar systems will be
    economically competitive with WCTs - but solar costs are
    highly sensitive to variations in solar irradiation -
    as shown in Figure 3.6.

3.5 Conclusion

We conclude from this discussion that, on economic
considerations, diesel, wind and animal powered systems are
the strongest candidates for economic analysis to establish
whether they offer better value for money for irrigation
applications than a water current turbine. Handpumps and
solar pumps are less promising alternatives. The remarks in
this section are intended to provide some initial guidelines
as to the circumstances in which each of the options may be
the most economically attractive. For further discussion see
reference 11 on p113.

COST vs HEAD

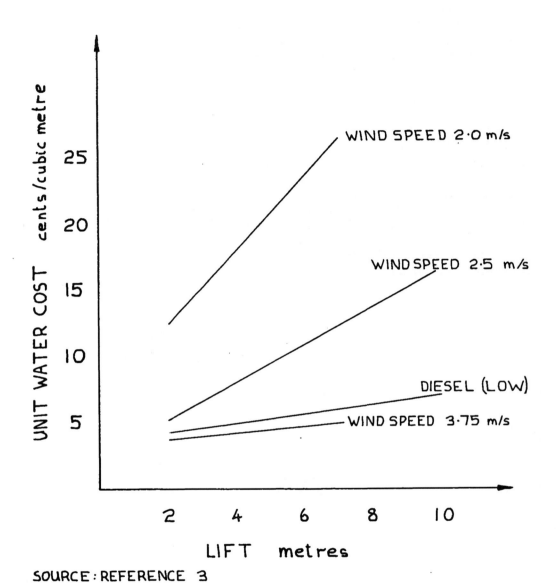

SOURCE: REFERENCE 3

FIGURE 3.5:   Sensitivity of Unit Water Costs of Wind Powered Systems to Wind Speed

COST vs HEAD

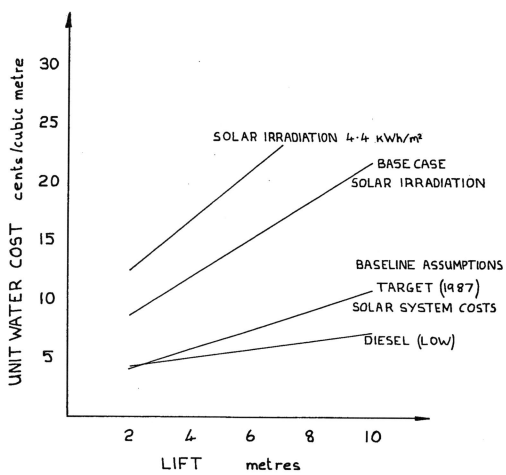

FIGURE 3.6: <u>Sensitivity of Unit Water Costs of Solar Powered Systems to Variations in Solar Irradiation.</u>

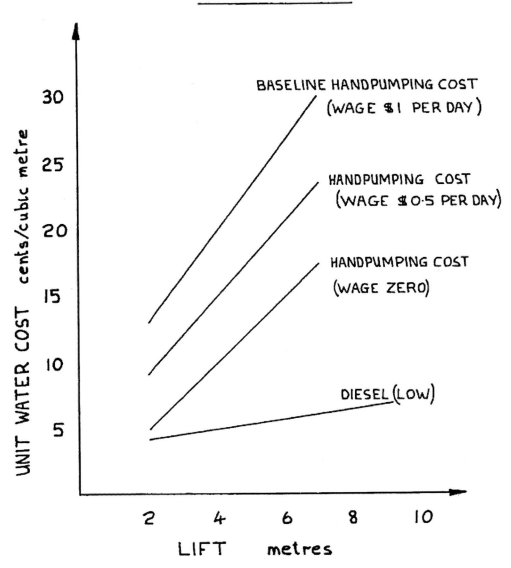

COST vs HEAD

BASELINE HANDPUMPING COST
(WAGE $1 PER DAY)

HANDPUMPING COST
(WAGE $0·5 PER DAY)

HANDPUMPING COST
(WAGE ZERO)

DIESEL (LOW)

UNIT WATER COST cents/cubic metre

LIFT metres

SOURCE: REFERENCE 3

FIGURE 3.7:   Sensitivity of Unit Water Costs of Handpump
Systems to Costs of Labour.

CHAPTER FOUR

Conclusion

This chapter provides a check list of the key social, economic and technical factors to be considered at the appraisal stage. The importance attached to each of these factors will vary enormously from place to place and in many places there may be additional important considerations to those indicated here.

4.1 Social Factors   See Section

1.   Do the farmers own or have traditional
     rights to the land to be irrigated?   3.3

2.   Are there other people who traditionally
     use this land who will be adversely
     affected?   3.3.5

3.   Will the presence of turbines in the river
     affect other river or river bank users?   3.3.5

4.   Is there any tradition of irrigated
     vegetable cultivation in the area or
     amongst the farmers?   2.15

5.   Have previous co-operative enterprises
     been successful enough to hope that
     several farmers could share one 'Mark 1'
     machine?   3.3.4

6.   Are there established mechanisms for the
     collection of money from the farmers for
     loan repayment?   3.3.4

4.2 Economic Factors

1.   Is there a large market for vegetables in
     the dry season?   3.2

2.   Is any form of pumped irrigation
     economically viable?   3.2

3.   Is WCT technology cheaper than alternative
     systems?   3.4

4.   What are the water requirements per hectare
     for dry season vegetable cultivation?   A.3.2

5.   Which distribution method is most
     economic?   2.8

6. Is a storage tank worthwhile?   2.13

7. Is sufficient labour available to work the
    scheme?

8. Is capital available for the purchase of
    turbines, pipes and tanks?

9. Is there access to foreign currency?   3.3.6

4.3 Technical Factors

4.3.1 Site Conditions

1. Is there sufficient water current speed
    when irrigation is required?   2.1.3

2. Is there sufficient water depth when
    irrigation is required?   2.1.4

3. Will the turbine interfere with river
    traffic?

4. How much weed and other debris is there
    in the river?   2.15.2

5. What delivery pipework is required?   2.8
                                                        A.1.3

6. What is the static pumping head required?   2.2

7. Is there a suitable site for a water
    storage tank (if required)?

8. Is a boat or canoe required?   2.15.3

9. Can the turbine be towed to site or is
    there a suitable place for its assembly?   2.14.2

10. Is theft of the turbine or pipework
    likely to be a problem?

4.3.2 Manufacturing

1. Are there people with the following
    skills: 3.3.6

    Metal turning
    fitting
    arc welding
    sheet metal work

carpentry and joinery
ferrocement (principally plastering)
engineering technician (Quality control,
calculation of transmission ratios, design
alterations to suit materials available,
technical problem solving)?

2. Are the following tools (or substitutes)
available:

Centre lathe
pillar drill
arc welder
hand tools for all skills above?

3. Are these (or similar) materials obtainable:

Mild steel sections (angle up to 50 mm, channel
up to 75 mm, flat plate up to 9 mm, rod up to 12 mm)
galvanized steel water pipe up to 75 mm
alkethene pipe up to 75 mm bore
steel cable, 8 mm diameter
fasteners (nuts, bolts, rivets, cable grips)
12 mm mesh chicken wire
cement
seasoned timber
oil drums ('Low Cost' machine)
paint
self aligning pillow block bearings
centrifugal pump, belts and pulleys (for 'Mark
1' machine)
cycle components (for 'Low Cost' Machine)?

4.3.3 Operation

1. Do the farmers have the necessary basic skills
to operate the machines and the irrigation
system?   2.15

2. Are the farmers sufficiently motivated to keep
the machine running?

3. Are extension workers available to provide
training in agricultural techniques?

4. What first aid and treatment facilities are
available in the event of an accident?

4.3.4 Maintenance

1. How much of the maintenance can be taken on
by the farmers?

                                                            2.16

2. How are they to be trained to carry out this maintenance?

3. Do they have access to the tools required for regular maintenance?

4. What arrangements will be made for maintenance and repair work beyond the farmers capabilities to be carried out?

5. How will the maintenance be paid for?

6. Will any guarantee be given on the turbine's performance or reliability?

APPENDIX ONE

More Detailed Technical Information

Al.1 Rotor Performance Comparisons

As already explained, the Darrieus and propeller type rotors operate on lift forces. The turbine blades have a hydrofoil cross section (see, for example, Figure Al.1) which, when it moves at an angle relative to the current direction, produces a lift force at right angles to the relative velocity of the water as seen from the blade.

Figure Al.1 shows how the relative velocity is found by vector addition of the stream and blade tip velocities. The ratio between these two velocities is known as the tip speed ratio, and is an important parameter used in setting the correct transmission ration.

$$\beta = \frac{u_B}{v_s} \quad \text{where the blade velocity } u_B = \frac{2\pi Nr}{60}$$

where N = rpm of rotor

where r = radius of rotor

The value of β depends on the type of rotor, the number of blades, and the load on it.

For a three-blade propeller type rotor with NACA 0025 (reference 16, p114) section blades running without any load connected is about 5.5. In other words, the blade tips are moving at 5.5 times the river speed. At $\beta_{\text{no load}}$ all the power developed by the rotor is dissipated by drag.

As can be seen from Figure Al.1 the lift force acting on the blade can be resolved into two components; parallel and normal to the plane of rotor rotation. The parallel component makes the rotor turn and the normal component bends the blade. The lift force is proportional to the angle of attack (α) up to the stall angle of the hydrofoil. As load (eg a pump or generator) is applied to the turbine, it slows down. This has the effect of increasing α (the angle between the blade chord and the V_R) and hence increasing the lift force. As further load is applied α increases, until eventually it exceeds the stall angle of the hydrofoil section and that part of the blade no longer contributes to the power output of the rotor. Once large areas of the blades are in the stalled condition, the turbine simply stops. Figure Al.2 shows the performance curves for a Darrieus rotor and a propeller rotor. These curves are equivalent to power versus rotational speed curves but by plotting C_p vs β the curves become independent of river speed and therefore more widely applicable.

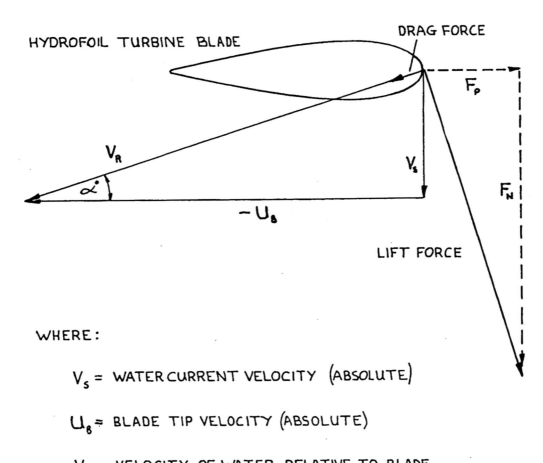

HYDROFOIL TURBINE BLADE

DRAG FORCE

F_P

V_R

α°

V_s

F_N

$-U_B$

LIFT FORCE

WHERE:

V_s = WATER CURRENT VELOCITY (ABSOLUTE)

U_B = BLADE TIP VELOCITY (ABSOLUTE)

V_R = VELOCITY OF WATER RELATIVE TO BLADE

F_P = COMPONENT OF LIFT FORCE IN
PLANE OF ROTOR ROTATION

F_N = COMPONENT OF LIFT FORCE NORMAL
TO PLANE OF ROTOR ROTATION

FIGURE A1.1: Rotor Blade Hydrodynamics.

C_p vs β CURVES

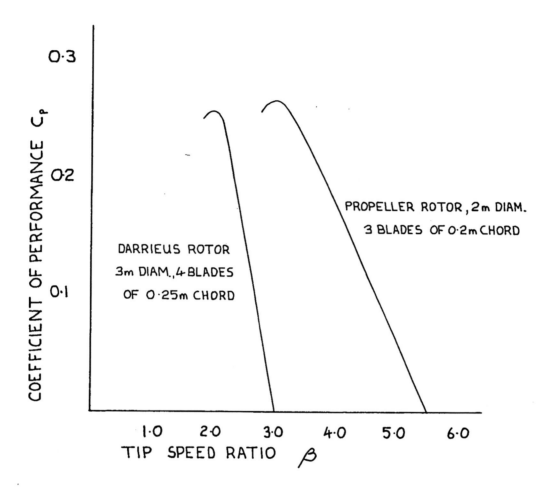

FIGURE A1.2: Performance Curves for Darrieus and Propeller Turbine Rotors.

From the curves it can be seen that the propeller turbine will run at tip speed ratios between 5.5 and 3, the Darrieus between β = 3 and β = 2. To obtain the maximum power output from a rotor it should be loaded until it is running as close to $\beta c_{p \ max}$ as possible (ie at β = 3 for the three-bladed propeller rotor).

Generally speaking, it is desirable to select as high speed a rotor as possible, because the faster the loaded rotor turns the cheaper will be the transmission. Reducing the number of blades or the blade chord length tends to increase the rotational speed, but smooth running and structural considerations set minimums for both these variables.

A1.2 Rotor Bearing Loadings

The rotor drag can be calculated if the rotor is treated as a flat disc and the change in velocity across it is assumed to be that which would give optimum performance, that is, a reduction of current speed of two thirds.

$$\text{Rotor Drag force, } D = A_s \ (2/3 \ V_s)^2 \ \ldots \ [5]$$

Measurements on actual machines have shown that for propeller type rotors this equation errs on the safe side (ie gives slightly too high a drag force) at current speeds of up to 1.25 m/s. For speeds between 1.25 m/s and 1.4 m/s the equation should be modified to:

$$\text{Rotor Drag force, } D = A_s \ (3/4 \ V_s)^2 \ \ldots [6]$$

No tests have been carried out at higher speeds than 1.4 m/s but for propeller rotors of less than 1 kW output power the rotor drag force is not expected to be greater than 3,500 N. This would produce an axial force on the rotor shaft top bearing of D cosΘ, (see Figure 2.7c) that is, about 2,700 N.

The radial load on this bearing will depend on the type of transmission used, but for the 'Mark I' machine the first stage belt tension exerts a sideways force of 2,400 N on the bearing.

A pillow block bearing which will give a life of at least 10 years continuous running with these loadings should be selected. Bearing manufacturers' technical literature shows how to convert a mixed axial and radial load into an equivalent radial load for the purposes of bearing selection.

Most self-aligning pillow block bearings have two grub screws on the inner housing, which provide location on the shaft. These grub screws are not strong enough to take the axial load on the rotor shaft which must therefore have a shoulder machined on it for the bearing inner to locate against. The bearing should be fitted to the shaft with the grub screws removed, their positions marked on the shaft so that it can be dimpled by drilling to let the grub screws get a good grip. This is necessary to avoid any chance of the bearing inner housing rotating relative to the shaft which will cause heating, loss of bearing grease and rapid failure of the bearing. This is particularly true of the intermediate shaft bearings with 10 times the rate of rotation. Whilst on the subject of bearings, it is important to note that the shaft diameter under the inner <u>must</u> be within the tolerance given by the bearing manufacturers. This is normally + 0.00 - 0.05 mm on the size of bearings we are dealing with. If the inners are loose on the shaft the problem of relative movement is more likely, and because the grub screws pull one end of the inner to one side, the plane of the bearing track will no longer be perpendicular to the axis of rotation of the shaft. Needless to say, this results in premature bearing failure. The bearing seats on the rotating shafts are the only parts on the 'Mark I' and 'Low Cost' machines where accurate turning is required, but it is critical that the tolerances called for are achieved to obtain reasonable bearing life.

A1.3 Calculation of Required Transmission Ratio

As mentioned in A1.1, to obtain the maximum rotor power (and hence maximum system efficiency and minimum capital cost per unit of output) it is necessary to load the three-bladed propeller rotor until it is running at a tip speed ratio of three. The load on the rotor is changed by altering the transmission ratio and the ratio required at a given site will vary depending on the static head, the head loss due to friction in the delivery pipe and the rotor diameter fitted to the machine. The optimum transmission ratio for a given site can be found by trial and error at the commissioning stage but much time and effort can be saved by calculating the required transmission ratio.

There are five stages in this procedure:

1. Estimation of head loss vs discharge curve for pipe system.

2. Calculation of pump performance at different rotational speeds.

3. Calculation of pump power requirements at different speeds and discharges.

4. Calculation of rotor speed and power available to drive pumps.

5. Comparing pump and rotor speeds at matching power levels to find required transmission ratio.

FRICTION LOSS vs DISCHARGE

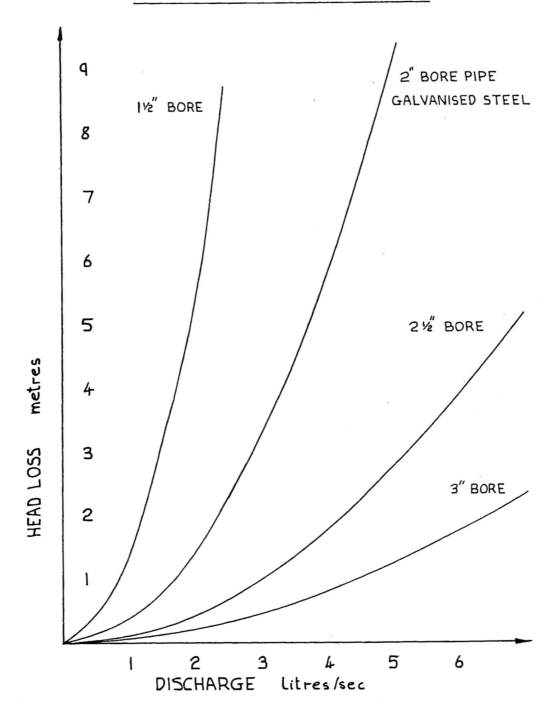

FIGURE A1.3: Pipe Friction Loss as a Function of Discharge for 60 metre Pipelines.

The following information is required: The static pumping head, details (length, diameter and material) of the delivery pipeline at the site, river speed and the manufacturer's 'head vs discharge' and 'power vs discharge' curves for the pump.

A1.3.1 Estimation of Head Loss vs Discharge Curve for Pipe System

The water discharged from the pump must be delivered to a convenient point on the river bank from which it can flow by gravity to the crops being irrigated. In Juba this typically involves a vertical distance or static head of 3-5 metres and a horizontal distance of 5-20 metres to the river bank, and then 10-40 metres to the end of a flexible pipe which is moved to deliver water to different earth channels which convey it to different crops (see Figure 2.17).

The pump must not only lift the water through the static head but must also force it through the delivery pipe against the frictional resistance which varies in direct proportion to the pipe length, in inverse proportion to the fifth power of the pipe diameter and in proportion to the square of the discharge. This frictional resistance is quoted as so many metres of friction head and, when added to the static head, gives the total or dynamic head which the pump is working against. To get the maximum water discharge possible into the field it is necessary to keep the pipeline efficiency (ie the static head as a percentage of the total head) to a maximum. It can be seen from the above that the pipe diameter is the major factor influencing pipeline efficiency. The result of using too narrow a pipe is illustrated by the following example: A 'Mark 1' type machine was installed at a smallholder settlement scheme pumping to a static head of 4.4 metres through a pipeline consisting of 3 metres of 40 mm fire hose (with internal end connectors of even smaller diameter), 23 metres of 40 mm galvanized steel water pipe and 30 metres of 50 mm galvanized water pipe. The discharge onto the field was measured as 2.58 l/s and the dynamic head at the pump was measured as 13.2 metres.

Hence the friction head was 8.8 metres and the pipeline efficiency only 33 per cent. Calculations showed that three quarters of the friction loss was in the 40 mm pipe, its end fittings and bends. Adding 50 metres of 50 mm bore flexible alkathene pipe only reduced the discharge by 0.15 l/s and reduced the pipe efficiency by only 2 per cent. If the entire delivery system had consisted of 100 metres of 50 mm bore alkathene pipe, the discharge would have been approximately 3.5 l/s and the pipeline efficiency 50 per cent. The economic effects of pipeline efficiency are discussed in reference 8, p113, where it is shown that it is worthwhile increasing the pipe diameter until the pipeline efficiency is between 80 per cent and 90 per cent. In the case discussed above this would result in a discharge of at least five litres/sec. (with the 40/13 pump).

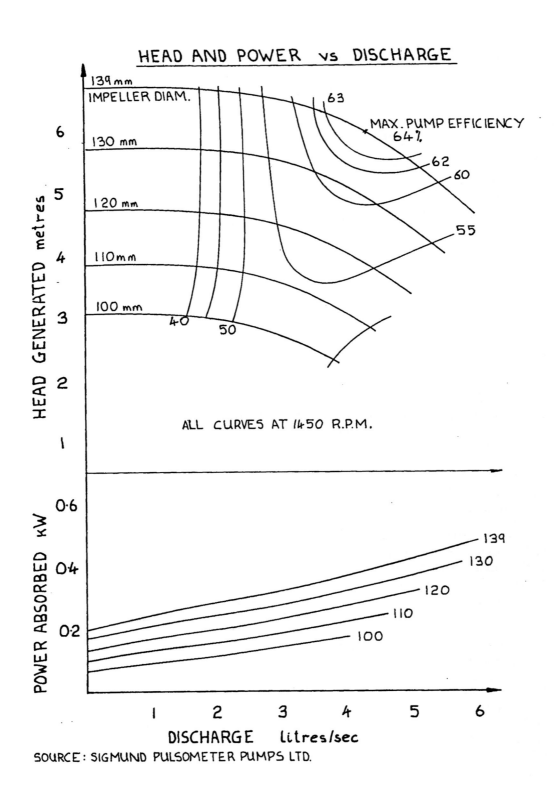

FIGURE A1.4: Performance Curve for SPP 40/13 Centrifugal Pump

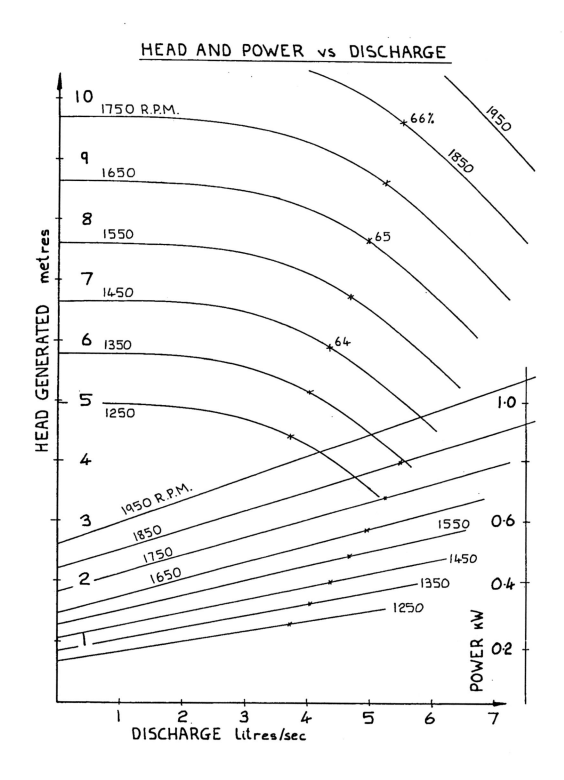

HEAD AND POWER vs DISCHARGE

FIGURE A1.5:   SPP 40/13 Pump Curves at Different Speeds.

93

Tables of friction losses in different diameters of pipe and losses in various types of bends and fittings are published by pump manufacturers and in text books (references 17 and 18, 114). For the purposes of pump selection it is convenient to plot a curve of head loss against discharge for the pipe system to be used at the site.

Figure A1.3 shows the pipe friction curve for different diameters of steel pipelines 60 metres long. To get the head loss for other lengths, simply multiply the head loss from the curve at the required discharge by the pipe length required and divide by 60. For smooth alkathene pipe divide the given head loss by 1.26. This factor corrects for the variation in surface roughness between the different materials.

Adding the friction head at a given discharge to the measured static head at the site gives the total head which the pump must generate at that discharge.

A1.3.2 Calculation of Pump Performance at Different Rotational Speeds

Dealing first with the 'Mark 1' machine, Figure A1.4 shows the head vs discharge and power vs discharge curves for the SPP 40/13 pump. The impeller specified for the 'Mark 1' machine is the 139 mm diameter, so in each case it is the top curve which is relevant. The top curve shows that at 1,450 rpm the pump has a maximum efficiency of 64 per cent and can generate a maximum head of 6.6 metres. The lower curve shows how the power absorbed by the pump increases with discharge.

Because the turbine shaft speed will vary with changing river speed it is necessary to plot the pump curves over the likely speed range. This can be accomplished using the affinity laws which state that:

(i)   the pump discharge is directly proportional to its rotational speed;

(ii) that the head generated is proportional to the square of the rotational speed; and

(iii) the power absorbed by the pump is proportional to the cube of its rotational speed.

The head vs discharge curves in Figure A1.5 are calculated for the different speeds by taking a series of points on the line at 1,450 rpm and using $Q \propto N$ and $H \propto N^2$ to calculate the corresponding H and Q valves at each of the other speeds.

For the 'Low Cost' machine no manufacturer's pump curve is available and because the efficiency obtained will depend on the quality of manufacture it is best to build a pump and test it at constant typical speed (say, driven at 1,500 rpm by an electric motor). The discharge at different heads can be measured by lifting a hose pipe (of at least 50 mm diameter and not more than 12 metres long to avoid friction losses spoiling the accuracy) up the side of a tall building or tree or, more simply, using a gate valve to restrict the pump output and a pressure gauge to measure the head developed by the pump. Figure A1.6 shows an approximate curve for the 'Low Cost' pump which has been estimated from the dimensions. This curve would be a reasonable starting point to find the required transmission ratio for a pump with a 150 mm impeller, 35 mm diameter inlet and 17 mm diffuser inlet diameter.

By the method already described, a family of head vs discharge curves for different speeds can be calculated. Note that for clarity the scale on the discharge axis of Figure A1.6 is twice that of A1.4,5 etc, and so Figure A1.3 must be replotted before it can be used with Figure A1.6.

A1.3.3 Calculation of Pump Power Requirements at Different Speeds and Discharges

The power vs discharge curve is slightly complicated by the fact that the efficiency of a centrifugal pump increases as its rotational speed goes up. Thus it is necessary to calculate the maximum efficiency at each pump speed using the Moody equation:

$$\frac{(1 - E)}{(1 - e)} = \frac{(n)}{(N)} 1/5$$

where E is the maximum efficiency at pump speed N and e is the efficiency at pump speed n

As can be seen from Figure A1.5 the actual variation in maximum efficiency in this case is quite small. The input power to the pump at the maximum efficiency point on each curve is calculated from the head, discharge and efficiency at each speed. (Input power to pump = pump efficiency x head x discharge x 9.81). At the zero discharge condition it is assumed that $P \propto N^3$ and hence the zero discharge power inputs could be found at each speed. A linear power variation with speed is assumed between the zero discharge power and the power at the maximum efficiency point on each curve, and hence the power vs discharge curve is plotted for each speed on the same sheet as the head vs discharge curve (see Figure A1.5).

A1.3.4 Calculation of Rotor Speed and Power Available to Drive Pump

The rotor diameter of the inclined axis machines can be increased up to 2.9 metres to enable a shaft power of 750 Watts to be produced at river speeds down to 1.1 m/s. In faster current speeds the turbine rotor's swept area is reduced by fitting shorter blades. The rotor shaft power is kept below 1 kW on the 'Mark I' machine (350 Watts on the 'Low Cost' machine) to avoid overloading the transmission.

HEAD AND POWER vs DISCHARGE

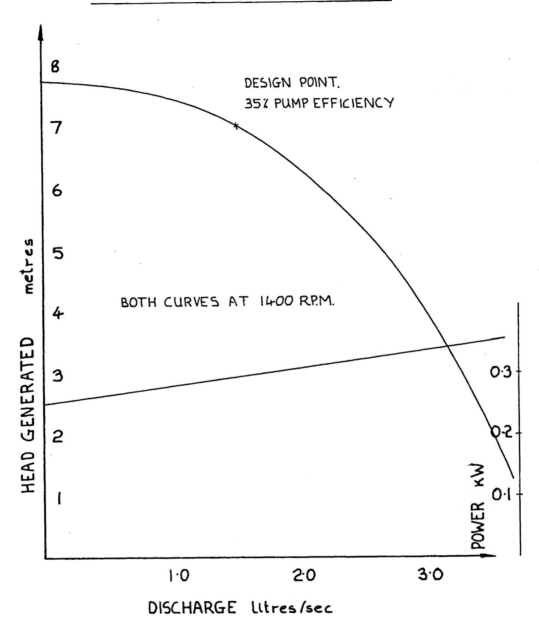

FIGURE A1.6: Performance Curve for Locally Fabricated Centrifugal Pump.

Three different blade lengths are detailed in the 'Mark 1' design and the rotor hub has two alternative positions for each blade, giving a total of six possible diameters as follows:

Rotor diameter	Blade size	Swept area at 40° inclination	River Speed for 1 kW shaft power	River Speed for 750 Watts shaft power	Shaft Speed at 750 Watts
m	m	m^2	m/s	m/s	rpm
2.9	1.45	5.06	1.16	1.06	20.9
2.71	1.45	4.42	1.22	1.11	23.5
2.5	1.25	3.76	1.29	1.17	26.8
2.31	1.25	3.21	1.36	1.23	30.5
2.0	1.0	2.41	1.49	1.36	39
1.81	1.0	1.97	1.6	1.45	45.9

Table A1.1   Rotor Sizes for 'Mark I'

From Table A1.1 it can be seen that the 1.45 metre blades should be used for any river speed up to 1.2 m/s. Between 1.2 and 1.4 m/s the 1.25 metre blades should be used and at speeds between 1.4 and 1.6 m/s the 1 metre blades are suitable. At river speeds above 1.6 m/s the 0.75 metre blades detailed in the 'Low Cost' design should be fitted.

In Table A1.1 the river speeds are calculated using equation 2 in Section 2.1.2, assuming that the rotor efficiency, C_p, is constant at a value of 0.25. From the limited performance test results of the inclined axis rotor this seems reasonable for well made aluminium alloy sheathed blades.

HEAD AND POWER vs DISCHARGE

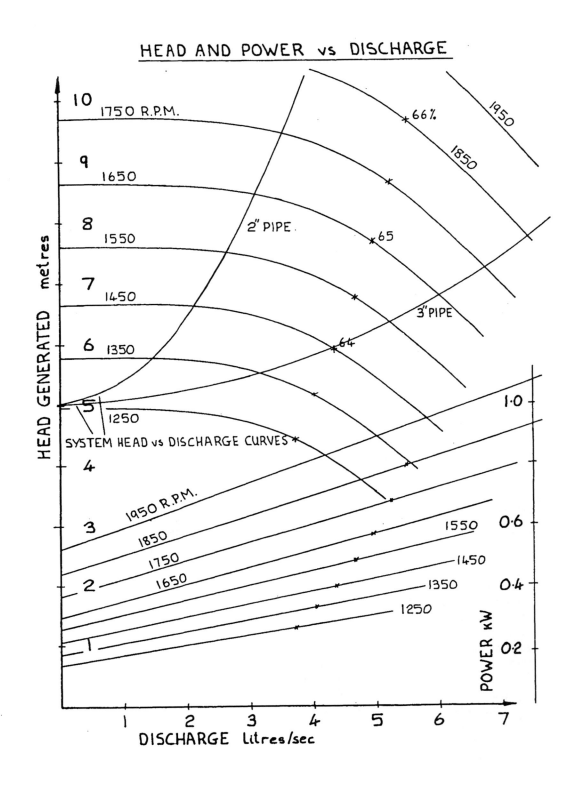

FIGURE A1.7: Head versus Discharge Curves for Two Possible Delivery Systems.

The efficiency of the Poly V belt transmission has been shown to be 90 per cent per stage and hence when the shaft power is 750 Watts about 600 Watts will arrive at the pump.

For the 'Low Cost' machine the blade lengths and shaft speeds are as follows:

Rotor diameter	Blade size	Swept area at 40° inclination	River speed for 350 Watts shaft power	River speed for 240 Watts shaft power	Shaftspeed at 240 Watts
m	m	m^2	m/s	m/s	rpm
2.5	1.25	3.76	0.91	0.8	18.3
2.31	1.25	3.21	0.96	0.84	20.8
2.0	1.0	3.41	1.05	0.93	26.6
1.81	1.0	1.97	1.12	1.0	31.7
1.5	0.75	1.35	1.28	1.12	42.8
1.31	0.75	1.03	1.4	1.23	53.8

Table A1.2   Rotor Sizes for 'Low Cost'

If it proves necessary to put the machine in a river speed of greater than 1.4 m/s a set of even shorter blades will have to be manufactured.

Testing has indicated that the low cost transmission has an efficiency of approximately 85 per cent so that at a river speed of 0.8 m/s approximately 200 watts is the power available to drive the pump.

A1.3.5 Comparing Pump and Rotor Speeds at Matching Power Levels to Find Required Transmission Ratio

By superimposing the relevant pipeline friction loss curve (those in Figure A1.3 are typical examples but you should plot your own delivery system curve and then transfer it onto tracing paper so that you can see your pumps curves through it) on the pump curves starting at the appropriate static head, the total dynamic head and discharge at a given pump

99

speed can be found at the intersection of the delivery system curve and the pump curve at that speed. Figure A1.7 shows two system curves for the 'Mark 1' machine, each with a static head of 5 metres and a 60 metre steel pipeline, but one using 2" bore pipe and the other 3". To find the power required at the various pump speeds, the intercepts are simply projected vertically down to the relevant power curve and hence a power vs discharge curve for each system is produced as in Figure A1.8. Given that the power available to drive the pump is 600 Watts, then the pump speed which would absorb 600 Watts can be found for each of our delivery systems from their respective power vs discharge curves. Referring to Figure A1.8 it can be seen that with the 2" bore pipe the pump should run at 1,775 rpm, the system discharge will be 3.6 litre/sec and the total head will be 9.6 metrtes, giving a pipeline efficiency of 53 per cent. With a 3" bore pipe the pump should run at 1,630 rpm, deliver 5.85 litres/sec and generate a head of 6.65 metres, giving a pipeline efficiency of 75 per cent. In both cases the pump is running within reasonable distance of its maximum efficiency at that speed.

If the river speed was, say, between 1.2 m/s and 1.3 m/s, then the 1.25 metres blades would be fitted so as to give a rotor diameter of 2.5 metres and a rotor speed of 26.8 rpm at 600 Watt pump inlet power. Hence for the 2" delivery pipe the required transmission ratio is 66.2:1 and for the 3" pipe 60.8:1.

Exactly the same procedure is followed in the case of the 'Low Cost' machine to find the required transmission ratio for any delivery system, static head and rotor diameter. However, the maximum speed the pump can be driven at with 200 Watts is 1,250 rpm; hence the maximum speed step up required for this machine is 68:1. This is achieved using a 48-tooth sprocket on the rotor shaft driving a 12-tooth sprocket via the chain and a 40 mm diameter on the pump shaft where it is in contact with the tyre. Higher ratios would be feasible if sprockets with more than 48 teeth could be obtained. The transmission ratio can be reduced by fitting a small sprocket with more teeth or preferably by increasing the pump shaft diameter.

If the river speed is below 1.09 m.s then even with the 2.9 metre rotor diameter the 'Mark 1' machine will not generate 750 Watts. To find the delivery if, for example, the river current speed was 0.85, Equation 2 in section 2.12 is used and the result multiplied by the transmission efficiency:

$$\text{Power at pump} = 1/2 \rho A_s v^3 C_p \times 0.8$$

$$= 311 \text{ Watts}$$

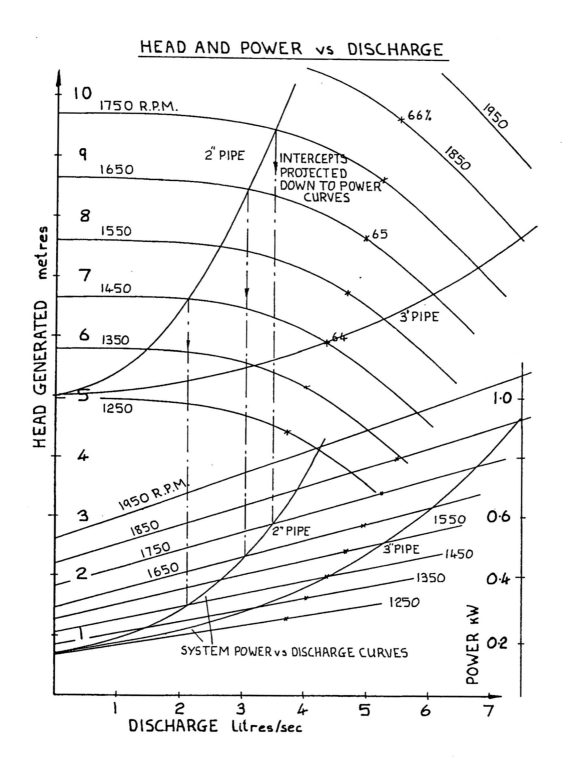

HEAD AND POWER vs DISCHARGE

FIGURE A1.8: Power Versus Discharge Curves for Two Possible
Delivery Systems.

Using this power level on Figure A1.7 we see that with the same static head and pipe system the discharge would be 2.2 l/s at a pump speed of 1,470 rpm with the 2" pipe and 3.45 l/s at 1,375 rpm with the 3" pipe. The correct rotor speed at 0.85 m/s is found using the tip speed ratio equation from Appendix 1.1 and hence at $\beta = 3$, the rotor should rotate at 16.8 rpm. The required transmission ratios are therefore 87.5 and 81.9 with the 2" and 3" delivery pipes, respectively.

In general terms, the higher the total head, the higher the transmission ratio and the larger the rotor diameter the higher the transmission ratio.

Having decided on the transmission ratio it is now possible to plot curves of discharge vs river speed from the Power vs Discharge curves in Figure A1.8. Figure A1.9 shows the curves for our two example systems across the river speed range 0-1.6 m/s. Note that the small additional capital cost of the three-inch delivery pipe would be well justified as it increases the water output by about 66 per cent.

Figure A1.9 may give the impression that frequent changes of rotor diameter (and hence transmission ratio) are required. In practice it is unlikely that the current speed will vary more than 10 or 15 per cent over the dry season and so for any given site the rotor diameter which will produce 1 KW at the <u>maximum</u> expected current speed is chosen. Running the machine at above 1 kW shaft power will produce rapid belt wear and at very high power levels failure the rotor hub or blades is likely.

A1.4 Estimation of Overall System Efficiency

The overall system efficiency mentioned in Section 2.1.4 is the ratio of the hydraulic power output from the delivery pipe to the power available in the water flowing through the turbine rotor. Section 2.14.3 describes the method of measuring the system efficiency at the commissioning phase, but it is also possible to calculate the probable efficiency before installation of the machine. If the efficiencies of the various elements of the design are known the overall efficiency is given by:

overall efficiency = rotor efficiency x transmission efficiency x pump efficiency x pipeline efficiency

$$\eta_o = C_p \times \eta_{Trans} \times \eta_{Pump} \times \eta_{Pipe}$$

typical figures for 'Mark 1' machine

$$C_p = 0.25$$

$$\eta_{Trans} = 0.8$$
$$\eta_{Pump} = 0.6$$
$$\eta_{Pipe} = 0.75$$
Hence $\eta_o = 9$ per cent

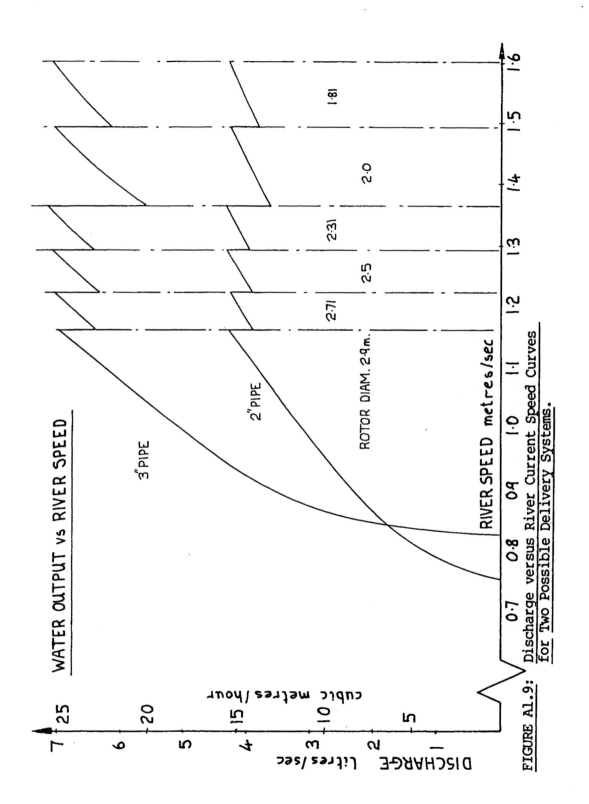

FIGURE A1.9: Discharge versus River Current Speed Curves for Two Possible Delivery Systems.

103

typical figures for 'Low Cost' machine

$$C_p = 0.25$$
$$\eta \text{ Trans} = 0.9$$
$$\eta \text{ Pump} = 0.35$$
$$\eta \text{ Pipe} = 0.85$$
$$\text{Hence } \eta_o = 6.7 \text{ per cent}$$

If some of these efficiencies can be measured on site (such as pipeline efficiency using a pressure gauge and overall system efficiency) then fault finding is often simplified.

A1.5 Maintenance Instructions for 'Mark 1' Water Pumping Turbine

Daily

Clear weed from the machine.
Clear weed and grass from pump inlet strainer.

Monthly

Check mooring cable fastenings and mooring post.
Check belts for tightness and correct position.
Check delivery hose connections.
Check all nuts and bolts for tightness on rotor, transmission and pontoon (17, 19 and 10 mm spanners).
Check condition of brake blocks. Replace if necessary.

Three-monthly

Check condition of ferrocement floats.
Check bottom rotor shaft bearing for wear. Replace pin or insert if necessary.
Check winch wires for corrosion.

Yearly

Grease rotor shaft top bearing and transmission bearings – four strokes each of a grease gun. Grease nipple nut must be screwed into bearing housing first.
Replace winch wires.
Check belts for wear and replace if necessary.

APPENDIX TWO

Case Study of the Economics of Using Water Current Turbines for Irrigating Vegetable Gardens in Juba

This case study has two objectives:

(i)   it illustrates the method of assessing whether WCTs are profitable technology to be used for irrigation: and

(ii) it demonstrates that WCTs are profitable for this purpose in Juba.

The results presented are intended primarily to be illustrative. Further details are available in references 1 and 2, on p113.

The steps in the calculation (which follow the method outlined Section 3.2) are set out in Table A2.1. The adopted measure of economic viability is that of the payback period. This is the length of time required to pay back capital costs, including interest, from crop revenues.

The caluclation in Table 2.1 shows that, on the assumptions made, the 'Low Cost' version of the WCT, used to grow vegetables during the dry season in southern Sudan, will pay for itself in one-and-a-half-dry seasons - ie in a total elapsed time of 15 months. In practice, an anaylysis should be conducted of the sensitivity of such a result to changes in the values of key input variables. Some of the critical variables are:

(i)   capital costs;

(ii) achievable water output;

(iii) crop yields;

(iv) existing crop prices;

(v)   the extent to which crop prices will fall in response to increases in supply due to the availability of irrigated crops grown in the dry season.

In this example, the available evidence (reference 2 on p113), particularly on crop yeilds and on the likely fall in prices, is somewhat uncertain. Crop prices during the rainy season can provide a useful guide to the lower limit to which dry season prices are likely to fall. The input price and revenue data in Table A2.1 is based on this indicator. However, Juba is an isolated town of some 100,000 people; crop prices, may, therefore, exceed those elsewhere throughout the year - making the economics of the turbine particularly favourable.

(i)   Input Data

WCT size = 'low cost' machine (2.5 m^2 swept area)
River current speed = 1 m/sec
Water lift   = 5 m
Water output   = 1 litre/sec
                  = 28.8 m^3 per eight hour day
                  = 5,200 m^3 per 180 day dry season
Irrigated area   = 1 feddan (0.4 hectares)

Estimated yields per feddan per season:
     Aubergine : 100,000
     Okra   : 160,000
     Kudra   : 10,000 bundles/month

Present dry season prices:
     aubergine 5p, okra 2.5p, kudra 5p/bundle

(ii)   Estimated production costs

WCT capital cost = S32,070

WCT running cost per season:
Labour : three workers @ S£30/worker/month = S£540
Spare parts : none   =   0

Other input costs:
     Seeds   = S£100
     Fertilizer   None   0

     TOTAL recurrent costs/season   = S£640

(iii)   Estimated irrigation benefits
At present dry season prices
     Aubergine = 100,000 at 5p   = S£5,000/feddan/season
     Okra   = 160,000 at 2.5p = S£4,000/feddan/season
     Kudra   = 10,000 at 5p   = S£ 500/feddan/month
                                   S£3,000/feddan/season

At revised[1] dry season prices
estimated revenue   S£2,000/feddan/season

TABLE A2.1 : Calculation of Benefits and Costs of Using a
             Water Current Turbine to Irrigate Vegetable
             Gardens in Juba.

[1] Conservative price estimate based on wet season prices

106

Table A2.1 continued

(iv)   <u>Estimated operating surplus per dry season</u>

equals:
    Estimated revenue   S£2,000
less:
    estimated recurrent costs   S£ 640
equals:
    operating surplus   S£1,360

(v)   <u>Estimated payback period</u>

equals:
    capital cost   S£2,070
divided by:
    operating surplus   S£1,360
equals: 1.52 dry seasons

That is, on the basis of these assumptions, the capital costs
of purchasing a water turbine will be paid for out of
addtional revenue from vegetable sales in one-and-a-half dry
seasons - a total elapsed time of 15 months.

APPENDIX THREE

Evidence on the Costs of Using Water Current Turbines for Irrigation Compared with the Costs of Using Alternative Pumping Methods

A3.1 Introduction

As explained in Section 3.4.1 in the main text, evidence on the costs of constructing and operating water current turbines is limited to that of the experience in southern Sudan (references 1 and 2, p113). This study compares WCTs to diesel pumps. In addition, a recent detailed study (reference 3, p113) has calculated the costs of irrigation pumping using solar, wind, diesel, animal and handpumps. By combining these two sources of information we can make some preliminary comments on the costs of using WCTs compared to alternative systems.

The alternative technologies are compared on the basis of unit water costs. The main steps in the estimation of these costs are:

    (i)   to specify the maximum volume of water output required per hectare - for systems not based on renewable energy, this will determine the size of system required;

    (ii)   to identify, for systems based on renewable energy sources, the month when the ratio of energy required to energy available is at a maximum - to determine the size of systems required;

    (iii)   to identify the capital, maintenance and operating costs associated with supplying water using each technology (over a specified analysis period) and the years in which these costs will be incurred;

    (iv)   to discount future costs, using a specified discount rate, to a common base year (this aggregate figure is known as the total discounted lifecycle cost); and

    (v)   to divide the total discounted lifecycle cost of each system by the total volume of water supplied during the analysis period to determine the unit cost (say, per cubic metre) of supplying water using each system.

A3.2 Comparison of Pumping Costs

A3.2.1 Evidence of Comparative Unit Water Costs from the Halcrow/IT Power Study (3)

The unit water costs of various alternative pumping systems for lifts of up to 10 metres estimated in reference 3, p112 are shown in Figure 3.4 of the main text. These costs relate to the 'baseline' irrigation case investigated; the main characteristics of this case are:

- (i)   irrigation area 2 hectares;

- (ii)   peak daily water requirement 60 cubic metres per hectare;

- (iii)   solar radiation in 'critical' solar month[1] 20.8MJ/square metre;

- (iv)   wind speed in 'critical' solar month 2.5m/second;

- (v)   discount rate – ten per cent;

- (vi)   analysis period equals 30 years.

The analysis is based on a host of assumptions on capital, installation, maintenance and operation costs and on energy efficiencies. Focusing on the performance of diesel-powered systems, the main conclusions to note are that:

---

[1]   The 'critical' month is that in which the ratio of energy requirement (ie water demand) to energy available is at at maximum.

(i)   diesel-powered pumping, based on 'low cost' assumptions[1], gives the cheapest unit water costs at all heights of lift, at a discount rate of ten per cent;

(ii)   wind power is cheaper than the diesel 'high cost' case[1] up to the lifts of nearly 10 metres; animal power is cheaper than 'high cost' diesel up to lifts of 7 metres;

(iii)   handpumps and solar pumps are more expensive than wind power and 'high cost' diesel.

There are, however, several important qualifications to these main conclusions. In particular:

(i)   unit water costs of irrigation using handpumps are sensitive to the value attached to rural labour; the 'baseline' assumption is a way of US$1 per day; if no value is attached to labour, unit water costs of handpumping fall to 5 and 17 cents per cubic metre at lifts of 2 and 7 metres respectively;

(ii)   solar and wind power systems (like WCTs) are developing technologies the capital costs of which are projected to fall - for example 'target' solar system costs (defined in reference 3, p112) are cheaper than all of the technologies shown in Figure 3.4 with the exception of 'diesel low'.

---

1   the diesel 'low' and 'high' cost assumptions are:

'low' : engine efficiency 15 per cent, engine life 5,000 operating hours or 10 years maximum, maintenance cost $200 per 1,000 operating hours, fuel price $0.4 per litre

'high' : engine efficiency 10 per cent, engine life 3,700 operating hours or seven-and-a-half years maximum, maintenance cost $300 per 1,000 operating hours, fuel price $0.8 per litre.
(All prices in US$ 1982.)

A3.2.2 Evidence on the Comparative costs of Pumping Using Water Current Turbines and Diesel Pumps in Southern Sudan

This study compared the costs of supplying water using the 'Mark 1' water current turbine and a 5-horsepower diesel pump. The main results are shown in Table A3.1. The key points to note are that:

(i) unit water costs using WCTs are highly sensitive to water current speed. This is because the power output is proportional to the cube of the current speed (see Section 2.1) and less supervision time is therefore required; and

(ii) diesel pumping costs are critically dependent on diesel prices; in remote regions of poor countries diesel is often in very short supply and only available at prices several times the official or subsidized price. This situation occurs in Southern Sudan - as evidenced by the unsubsidized price in Table A3.1.

From this evidence it can be concluded that from the farmer's viewpoint at water current speeds greater than or equal to 1.2 m/sec, WCTs are economically attractive compared to diesel pumps and, conversely, as speeds fall below 1 m/sec, diesel pumps become increasingly favourable. If water current speeds are between 1 and 1.2 m/sec, a careful analysis is essential.

A3.3 Conclusion

The evidence in the previous section has shown first that under specified 'baseline' conditions, diesel pumps are cheaper than the alternative pumping methods - particularly handpumps and systems based on solar power and second that, in the Southern Sudan, even at controlled diesel prices (which do not reflect the true scarcity of diesel), at current speeds in excess of 1.2 m/sec WCTs are cost-competitive with diesel.

We conclude from this evidence that water current turbines merit serious consideration as a cost-effective pumping technology if current speeds exceed 1.2 m/sec and may be the cheapest option at speeds over 1.0 m.sec.

Technology	Unit water costs S£/cubic metre
Water Current Turbine	
Water Current Speed (m/sec)	
0.8	0.34
0.9	0.24
1.0	0.17
1.1	0.13
1.2	0.1
1.3	0.08
Diesel	
Fuel Price	
Subsidized[1] S£0.5/litre	0.1
Unsubsidized[2] S£2.9/litre	0.27

Notes

(1) [1]   The subsidized price is the controlled price at which farmers should be able to obtain diesel.

(2) [2]   The unsubsidized price is the ruling 'free market' price at which farmers could obtain diesel at the time of the study.

(3)   Both prices assumed not to increase in real terms.

(4)   The costs are in units of S£ to prevent direct comparison with Figure 3.4. Information sources have made different assumptions and these data are not directly comparable.
(At February 1982 S£1 = US$1.1.)

(5)   Discount rate – ten per cent.

TABLE A3.1: Comparative Unit Costs of Water Pumping in Southern Sudan

REFERENCES

1   Garman, P (1983) 'Report to the Royal Netherlands Government on the design, construction and testing of low cost, river powered, water pumping turbines in Southern Sudan'. ITDG.

2   Marshall, K (1982) 'Evaluation of water current turbines in Southern Sudan'. ITIS.

3   (1983) 'Small-scale solar powered pumping systems the technology, its economics and advancement.' Sir William Halcrow & Partners in association with Intermediate Technology Power Ltd.

4   Burgess, Peter and Prynn, Peter (1985) Solar Pumping in the Future: A Socio-economic Assessment.   CSP Publications.

5   Kennedy, W K and Rogers, T A (1985) Human and Animal-Powered Water-Lifting Devices. ITDG.

6   Golding, E W (1955) The Generation of Electricity by Wind Power E and F N Spoon Ltd.

7   Garman, P (1982) 'Final Report to the Royal Netherlands Government on the overseas testing of the prototype river current turbine'. ITDG.

8   Garman, P (1983) 'The application of the vertical axis water turbine in developing countries'. M Phil thesis, Reading University.

9   Scout Farm Project (1984) The stream-driven coil pumps Parts I, II and III.   Danish Guide and Scout Association. Available from IT Publications.

10   Watt S B (1978)   Ferrocement water tanks and their construction. IT Publications.

11   Kenna, J and Gillet, B (1985) Solar Water Pumping.   IT Publications.

12   Gray, C and Martens, A (1983) 'The political economy of the 'recurrent cost problem' in the West African sahel'. World Development, II, 2.

13   'Bangladesh small-scale irrigation' (1983) USAID Project impact evaluation report no 42.

14   Chambers, R, Longhurst, R and Pacey, A (1981) Seasonal Dimensions to Rural Poverty. Frances Pinter Ltd.

15 Ruthenburg, H (1980) <u>Farming Systems in the Tropics</u>.
    Clarendon Press, Oxford: 3rd edition.

16 Abbot, H and Von Doenhoff, A (1958) <u>Theory of Wing
    Sections</u>. Dover Publications.

17 'Friction in Pipes' Bulletin No FC1577. Girdlestone
    Pumps Ltd.

18 'Flow in Pipes'. <u>Kempe's Engineers Handbook</u>.

19 Guerrero, Haime Lobo (forthcoming) 'Development of an
    efficient current water wheel'. Universidad de los
    Andes, Colombia. Paper to be published by ASME.